SECRETS AND KNOWLEDGE IN MEDICINE AND SCIENCE, 1500–1800

The History of Medicine in Context

Series Editors: Andrew Cunningham and Ole Peter Grell

Department of History and Philosophy of Science
University of Cambridge

Department of History
Open University

Titles in this series include

Henri de Rothschild, 1872–1947
Medicine and Theater
Harry W. Paul

The Anatomist Anatomis'd
An Experimental Discipline in Enlightenment Europe
Andrew Cunningham

Centres of Medical Excellence?
Medical Travel and Education in Europe, 1500–1789
Edited by Ole Peter Grell, Andrew Cunningham and Jon Arrizabalaga

Ireland and Medicine in the Seventeenth and Eighteenth Centuries
Edited by James Kelly and Fiona Clark

Negotiating the French Pox in Early Modern Germany
Claudia Stein

Before My Helpless Sight
Suffering, Dying and Military Medicine on the Western Front, 1914–1918
Leo van Bergen

Secrets and Knowledge in Medicine and Science, 1500–1800

Edited by

ELAINE LEONG

University of Cambridge, UK

and

ALISHA RANKIN

Tufts University, USA

Routledge
Taylor & Francis Group
LONDON AND NEW YORK

First published in paperback 2024

First published 2011 by Ashgate Publishing

First published 2016
by Routledge
4 Park Square, Milton Park, Abingdon, Oxon OX14 4RN

and by Routledge
605 Third Avenue, New York, NY 10158

Routledge is an imprint of the Taylor & Francis Group, an informa business

Copyright © 2011, 2016, 2024 Elaine Leong and Alisha Rankin

The right of Elaine Leong and Alisha Rankin to be identified as the authors of the editorial material, and of the authors for their individual chapters, has been asserted in accordance with sections 77 and 78 of the Copyright, Designs and Patents Act 1988.

All rights reserved. No part of this book may be reprinted or reproduced or utilised in any form or by any electronic, mechanical, or other means, now known or hereafter invented, including photocopying and recording, or in any information storage or retrieval system, without permission in writing from the publishers.

Trademark notice: Product or corporate names may be trademarks or registered trademarks, and are used only for identification and explanation without intent to infringe.

Publisher's Note
The publisher has gone to great lengths to ensure the quality of this reprint but points out that some imperfections in the original copies may be apparent.

British Library Cataloguing in Publication Data
Secrets and knowledge in medicine and science, 1500-1800.
-- (The history of medicine in context)
1. Medicine--Europe--History--16th century. 2. Medicine--Europe --History--17th century. 3. Medicine--Europe--History--18th century. 4. Discoveries in science--Europe--History--16th century. 5. Discoveries in science--Europe--History--17th century. 6. Discoveries in science--Europe--History--18th century. 7. Secrecy--Europe--History--16th century. 8. Secrecy--Europe--History--17th century. 9. Secrecy--Europe--History --18th century.
I. Series II. Leong, Elaine Yuen Tien, 1975- III. Rankin, Alisha Michelle.
610.9'4'0903-dc22

Library of Congress Cataloging-in-Publication Data
Secrets and knowledge in medicine and science, 1500-1800 / [edited by] Elaine Leong and Alisha Rankin.
p. cm.
Includes bibliographical references and index.
ISBN 978-0-7546-6854-1 (hbk)
1. Medicine--Europe--Philosophy--History. 2. Science--Europe--Philosophy--History. 3. Medicine, Magic, mystic, and spagiric--Europe--History. 4. Secret (Philosophy) I. Leong, Elaine Yuen Tien, 1975- II. Rankin, Alisha Michelle.
R484.S43 2011
610.94--dc22

2010049695

ISBN: 978-1-032-92473-1 (pbk)
ISBN: 978-0-754-66854-1 (hbk)
ISBN: 978-1-315-60800-6 (ebk)

DOI: 10.4324/9781315608006

Contents

PART IV SECRETS AND HEALTH

Notes on Contributors

Montserrat Cabré is Associate Professor of the History of Science at the Universidad de Cantabria, Spain. Her research interests include the history of medieval and early modern women's healthcare, medical conceptions of the female body as well as cultural representations of sexual difference. Her work has focused on late medieval and Renaissance constructions of the gendered body through medical discourse and practical experience; as a related issue, she has worked on the significance and types of bodywork provided by women in the household setting. She is currently completing a book on women's self-care in late medieval Iberia.

Sandra Cavallo is Professor of Early Modern History at Royal Holloway, University of London and specializes in the history of medicine, gender and material culture. She is the author of *Charity and Power in Early Modern Italy* (Cambridge University Press, 1995), *Artisans of the Body in Early Modern Italy: Identities, Families, Masculinities* (Manchester University Press, 2007), and the co-editor of *Widowhood in Medieval and Early Modern Europe* (Longman, 1999), *Spaces, Objects and Identities in Early Modern Italian Medicine* (Blackwell, 2008), *Domestic Institutional Interiors in Early Modern Europe* (Ashgate, 2009) and *A Cultural History of Childhood and the Family vol. 3, The Early Modern Age* (Berg, 2010). Her current research project looks at the construction of the healthy domestic environment and the healthy body in Renaissance and early modern Italy.

Michelle DiMeo recently completed an interdisciplinary PhD in English and History at the University of Warwick. Her thesis concerns the scientific and medical manuscripts of Katherine Jones, Lady Ranelagh (1615–1691). She has published essays on early modern women and medicine, contributed to the *Oxford DNB*, and is currently working on the collection of essays *Reading and Writing Recipe Books, 1600–1800* (co-edited with Sara Pennell).

William Eamon is Regents Professor of History and Dean of the Honors College at New Mexico State University. His research focuses on the history of science and medicine in Renaissance Italy and Spain, and on science and popular culture in early modern Europe. He is the author of *Science and the Secrets of Nature: Books of Secrets in Medieval and Early Modern Europe* (Princeton, 1994), *The Professor of Secrets: Mystery, Medicine, and Alchemy in Renaissance Italy* (Washington, 2010), and over 50 articles and book chapters on various aspects of early modern science and medicine. He is also the co-editor (with Victor Navarro Bròtons) of *Más allá de la Leyenda Negra: España y la Revolución Científica* (Valencia, 2007). He is currently at work on two book projects: *Science and Everyday Life in Early Modern Europe, 1500–1750* (Cambridge), and *Discovery and the Origins of Science.*

Michael Hunter is Professor of History at Birkbeck, University of London. He has written or edited many books on the history of ideas and their context in late seventeenth-century Britain, including a biography of Robert Boyle. He is also the Principal Editor of *Boyle's Works* (14 vols., 1999–2000), *Correspondence* (6 vols., 2000) and workdiaries (www.livesandletters.ac.uk/wd/index.html). From 2006 to 2009 he was Director of the AHRC-funded project, 'British Printed Images to 1700' (www.bpi1700.org.uk), and he edited the volume *Printed Images in Early Modern Britain: Essays in Interpretation* (Ashgate, 2010).

Elaine Leong is a Wellcome Trust Research Fellow at the Department of History and Philosophy of Science, University of Cambridge. Previously, she was a Leverhulme Early Career Fellow at the Department of History, University of Warwick and a visiting lecturer in the Department History and Philosophy of Science at the University of Cambridge. She has published articles on medicinal recipes and domestic medicine and is currently completing a monograph titled *Treasures for Health: Medical Knowledge and Practice in the Early Modern Household.*

Ayesha Mukherjee is Lecturer in Renaissance Literature and Culture at the Department of English, University of Exeter. Her PhD dissertation (University of Cambridge, 2007) was on the works of Sir Hugh Platt (1552–1608). She is currently turning this into a monograph titled *Penury into Plenty: Dearth and the Making of Knowledge in Early Modern England.*

Tara Nummedal is Associate Professor of History at Brown University, where she teaches courses on early modern Europe and the history of science. She is the author of *Alchemy and Authority in the Holy Roman Empire* (University of Chicago Press, 2007). Her current book project, *The Lion's Blood: Alchemy, Apocalypse, and Gender in Reformation Europe*, examines the intersection of gender and apocalypticism in the life of the sixteenth-century German alchemist Anna Maria Zieglerin.

Alisha Rankin is Assistant Professor of History at Tufts University, where she teaches early modern history and the history of science and medicine. Previously she was a Junior Research Fellow at Trinity College, University of Cambridge. She has published several articles on gentlewomen's medicine and empirical healers, particularly as related to the history of pharmacy. Her monograph, *Panacea's Daughters: Noblewomen and the Art of Healing in Early Modern Germany*, is currently under review, and she is working on a second book, *Wonder Drugs and Popular Pharmacy in Early Modern Germany*.

Lisa Wynne Smith is an Associate Professor at the University of Saskatchewan. She is working on two projects, a monograph titled *Domestic Medicine: Gender, Health and the Household in Eighteenth-Century England and France* and an online database of medical and scientific letters (Sir Hans Sloane's Correspondence Online). The Social Sciences and Humanities Research Council supported her research.

Pamela H. Smith is a Professor of History at Columbia University where she teaches history of early modern Europe and the history of science. She is the author of two award-winning books, *The Business of Alchemy: Science and Culture in the Holy Roman Empire* (Princeton, 1994) and *The Body of the Artisan: Art and Experience in the Scientific Revolution* (Chicago, 2004). She co-edited *Merchants and Marvels: Commerce, Science, and Art in Early Modern Europe* (with Paula Findlen, Routledge, 2002) and *Making Knowledge in Early Modern Europe: Practices, Objects, and Texts, 1400–1800* (with Benjamin Schmidt, Chicago, 2008). Her current book project is entitled *Making and Knowing: The Reconstruction of Historical Experience.*

Tessa Storey is Research Assistant on a Wellcome Trust funded project, 'Domestic Culture and the Prevention of Disease in Renaissance and Early Modern Italy' at Royal Holloway College, University of London. Prior to this she was Research Associate at Leicester University collaborating on a database of Italian Renaissance 'Books of Secrets'. In 1999 she held a Leverhulme Special Research Fellowship at Royal Holloway College. She has published the book *Carnal Commerce in Counter Reformation Rome* (Cambridge, 2008). Her other publications focus on the material culture of prostitution in seventeenth-century Rome and on the identity narratives of prostitutes.

Acknowledgements

This volume began life as a 2008 symposium held at the Centre for the Research in Arts, Social Sciences and Humanities (CRASSH) at the University of Cambridge. The meeting and the collaborative work for the volume was funded by the Wellcome Trust, the British Academy, the Leverhulme Trust, CRASSH, the Department of the History and Philosophy of Science and Trinity College at the University of Cambridge. We could not have organized the meeting without the administrative help provided by CRASSH, and we particularly extend our thanks to Michelle Maciejewska for her expert assistance from start to finish. Thanks also to Lauren Kassell and David Haycock for their input, advice and support at the early stages of the project. Finally, we are grateful to all the participants at the meeting for their contributions and commentaries.

As editors, we would like to thank all our contributors for our intellectually stimulating and enlightening 'virtual' conversations over the last year. It has been a pleasure to work with such a diverse group of scholars who challenged and extended our views of early modern knowledge making. We are also grateful to our colleagues at Cambridge, Tufts and Warwick for their support, suggestions and general camaraderie. Emily Yates at Ashgate has guided this project from start to finish with patience, grace and enthusiasm. Finally, we are grateful to our families for their unending patience and support.

Acknowledgements

Introduction

Secrets and Knowledge

Elaine Leong and Alisha Rankin

In 1555, Venetian humanist Girolamo Ruscelli published a book that influenced the world of vernacular print for decades to come. Printed pseudonymously under the name Alessio Piemontese, the book, *I secreti*, promised to reveal the author's long-kept 'secrets' and presented hundreds of recipes for the production of a wide range of substances, from medicines for the plague and pox to scented powders for perfuming the body to ink to counterfeit diamonds. It was an instant runaway success. Reprinted 17 times in its original Italian in the second half of the sixteenth century alone, the work was quickly translated into Latin, French, English, German, Dutch, Spanish and Polish. It also started a major publishing trend, with printers commissioning their own books of secrets.[1] As copies of *I secreti* and similar works rolled off European presses at an astounding rate, readers from all walks of life poured over the knowledge contained within. Copious marginal annotations can be found in surviving copies of every edition of *I secreti*, and many readers copied sections by hand into their personal notebooks.[2] *I secreti* and related works undoubtedly had a major impact on concepts of knowledge in early modern Europe.

Although she had never read *I secreti*, the Electress Anna of Saxony, a German noblewoman, had her own secrets. Around the time Ruscelli published his blockbuster work, Anna began a concerted effort to learn how to compound medicinal remedies. She sought out recipes much like the ones published under Alessio Piemontese's name, and she had acquaintances show her how to make

---

1 William Eamon, *Science and the Secrets of Nature: Books of Secrets in Medieval and Early Modern Culture* (Princeton, NJ: Princeton University Press, 1994), pp. 136 and 139.

2 As William Eamon argues in his chapter in this volume, many of the copies of William Ward's translation of *I secreti* are annotated. For example, the 1558, 1559, [1560], 1562, 1563, 1566, 1568, 1569, 1578, 1580 and 1615 editions of the text are all marked-up in some way: Folger STC 293 [1558 edition] copy 1 and 2, 295 [1559 edition], 300, 296 [1562 edition] copy 1 and 2 (bound with 301 [1562 edition, *Second Part of the Secrets of Maister Alexis*] and 305 [1562 edition, *Thyrd and Last Parte of the Secretes of the Reuerende Maister Alexis of Piemont*] copy 1 and 2), 297 [1568 edition] copy 1 and 2 (bound with 302 [1568? edition, *Seconde Parte*], 306 [1566 edition, *Thyrde and Last Parte*] and 309 [1569 edition] copy 1 and 2), 307 [1578 edition, *Third and Last Part*], 303 [1580 edition, *Seconde Part*] and 299 [1615 edition] copy 1 and 2. In addition, a lengthy 'digest' of an English edition of *I Secreti* can be found in the notebook containing the diary of the mid-seventeenth century non-conformist minister Thomas Larkham of Cornwall. British Library, Loan MS 9.

them. By the time of her death, her library contained over 50 medical manuscripts detailing cures similar to the medical recipes in Ruscelli's work.[3] Once acquired, she shared her knowledge with others – including the Duchess Anna of Bavaria, whom she befriended at the imperial election in 1562. In a long letter written in January 1563, she gave the duchess copious advice on making remedies and promised to send a servant to help her set up 'distilling and burning equipment'. She also enclosed several recipes and sent one ingredient – dog fat – that she feared the duchess might lack. Near the end of the letter however, Electress Anna expressed concern that she had revealed too much before receiving something in return. 'Now that we have trustingly given Your Dearest nearly all of our most secret arts', she wrote, 'we ask that … if Your Dearest knows of anything particular and special, that she not keep it from us … and we will preserve it for ourselves alone in all secrecy'. The knowledge she had divulged, her request intimated, was valuable and sensitive enough to deserve a *quid pro quo*. She added a plea that 'Your Dearest not make the [secrets] she has received from us common'.[4]

Nearly a century later, the London-based 'expert operator' Edward Fountain claimed to have 'the best secret in the World, absolutely and infallibly, to make Teeth most perfectly white and clean, though they be ever so black and rusty'. This secret appeared at the end of his 1649 publication, *A Pretious Treasury of Twenty Rare Secrets*, a curious hybrid of a book of secrets and an advertisement for his dental practice. The greater part of the *Treasury* reads much like the secrets of Alessio Piemontese. Here, we find instructions to 'make wine odoriferous' or to 'make a Candle burne in water, and to endure a long time', to cure the ague and to clean your feet from sweat. At the end, however, Fountain offered his readers a series of enticing dental services, such as the setting of artificial teeth, the making of 'smooth and plaine teeth', and the 'best secret in the World' for whitening teeth, all of which were offered by Fountain himself at his lodgings at St Martin's Lane.[5] *A Pretious Treasury*, thus, served a dual purpose of both establishing Fountain's position as a knower of secrets and natural knowledge and as a skilled dental practitioner. The work seemed to have been a success, as one year later, Fountain, now moved to the Strand near Temple-bar, issued another similar hybrid tract, this time titled *A Brief Collection of Many Rare Secrets … Very Necessary for All Sorts of People*.[6]

---

[3] Sächsische Landes- und Universitätsbibliothek, Dresden, MS K 342.

[4] Anna of Saxony to Anna of Bavaria, 5 January, 1563, Sächsische Hauptstaatsarchiv, Dresden, Kop. 511, fols 68r-69v (fol. 69r). 'Weil wir EL auch fast alle vnsere geheimbste kunste vortraulich mittgetheilt So bitten wir hinwider wo EL etwas gewisses vnd sonderliches hetten/sie woltenn vns dasselbige auch Schwesterlich nicht vorhalten/Solchs wollenn wir jnn gutter gehaimb bi vnnß bewaren/Jnnmassen wir vnß wunschen EL das jrige so sie von vnnß bekommen/auch nicht gemein machen werden'.

[5] Edward Fountain (or La Fountain), *A Pretious Treasury of Twenty Rare Secrets* (London, 1649), sig. A4v.

[6] London, 1650.

The fictitious Alessio Piemontese, the Electress Anna of Saxony and the 'expert operator' Edward Fountain, three figures inhabiting vastly different worlds in early modern Europe, were united by secrets and the knowledge they promised. In the intervening centuries, Alessio has become almost synonymous with books of secrets among early modern scholars, while Electress Anna and 'expert operator' Fountain are less well known in that capacity. Yet the fascination with collecting, evaluating and authoring medical and scientific secrets attracted diverse enthusiasts of all regions and walks of life – from Venetian humanists to London tooth-drawers, German noblewomen to French physicians, Spanish courtiers to the well-known Hartlib circle and Robert Boyle. For countless individuals, secrets held the key to unlocking the mysteries of nature, curing disease, maintaining good health, making practical everyday substances, and even creating wondrous tricks. Hunting for secrets was one of the main avenues through which early modern men and women attempted to satisfy their desire to understand the natural world around them. Secrets, whether presented in the form of a printed tract or a gathering of loose handwritten leafs, encompassed a large genre and a wide range of actors.

Although the main body of scholarship on secrets has been developed in the last 15 years, the study of books of secrets has its roots in the nineteenth century. In the early 1880s, John Ferguson (1838–1916), a bibliographer and chemist from the University of Glasgow, began to compile a series of bibliographical notes on 'books of technical receipts or so-called "secrets"', which he presented in a series of lectures and publications between 1882–1915. His painstaking and detailed work identified the genre and its importance, but it did not capture the attention of scholars for nearly a century.[7] The recent surge of historical interest has been spurred on in large part by the publication of William Eamon's groundbreaking *Science and the Secrets of Nature: Books of Secrets in Early Modern Culture* in 1995. Building on Ferguson's work, Eamon described the deluge of books of secrets issued by printing presses all over Europe and highlighted the sheer volume of literature overlooked by historians. Unlike Ferguson, moreover, he explained the significance of these works within their cultural and intellectual contexts. *Science and the Secrets of Nature* was thus the first monograph to engage with secrets as a topic of research and scholarship in their own right.

Since the appearance of Eamon's work, a number of scholars have explored medical, scientific and technical secrets from different angles. Most significant to our present volume is research in five overlapping bodies of scholarship. Firstly, a number of works have investigated the drive to uncover the 'secrets of nature' in early modern Europe, be they alchemical, astrological, botanical or medical.[8]

---

[7] John Ferguson, *Bibliographical Notes on the Histories of Inventions and Books of Secrets*, with a preface by William Eamon (Staten Island, NY: Pober Publishing, 1998).

[8] See, for example, Allen Debus and Michael T. Walton (eds), *Reading the Book of Nature: The Other Side of the Scientific Revolution* (Kirksville, MO: Sixteenth Century Journal Publishers, 1998); Lawrence M. Principe, *The Aspiring Adept: Robert Boyle and*

Also of great relevance is the movement focusing on the craft traditions, led by Pamela O. Long and Pamela H. Smith, which examines the role of artisanal know-how in uncovering nature's mysteries and the role of secrecy and openness within those traditions.[9] The secrets of nature and the drive towards scientific know-how have also been emphasized in works exploring the empirical study of nature, particularly in recent monographs by Brian Ogilvie and Antonio Barrera-Osorio.[10] Fourthly, the burgeoning literature on recipe books explores the fragments of natural knowledge that were collected, compiled and tested within the household.[11]

---

his Alchemical Quest (Princeton, NJ: Princeton University Press, 1998); Simon Varey, Rafael Chabrán, and Dora B. Weiner (eds), *Searching for the Secrets of Nature: The Life and Works of Dr. Francisco Hernández* (Stanford, CA: Stanford University Press, 2000); William R. Newman and Anthony Grafton (eds), *Secrets of Nature: Astrology and Alchemy in Early Modern Europe* (Cambridge, MA: MIT Press, 2001); and William R. Newman and Lawrence M. Principe, *Alchemy Tried in the Fire: Starkey, Boyle, and the Fate of Helmontian Chymistry* (Chicago, IL: University of Chicago Press, 2002).

9 Pamela Long, *Openness, Secrecy, Authorship: Technical Arts and the Culture of Knowledge from Antiquity to the Renaissance* (Baltimore, MD: Johns Hopkins University Press, 2001); Long, 'Power, Patronage, and the Authorship of Ars: From Mechanical Know-How to Mechanical Knowledge in the Last Scribal Age', *Isis*, 88/1 (1997): pp. 1–41; Pamela H. Smith, *The Body of the Artisan: Art and Experience in the Scientific Revolution* (Chicago, IL: University of Chicago Press, 2004).

10 Brian Ogilvie, *The Science of Describing: Natural History in Renaissance Europe* (Chicago, IL: University of Chicago Press, 2006); and Antonio Barrera-Osorio, *Experiencing Nature: The Spanish American Empire and the Early Scientific Revolution* (Austin, TX: University of Texas Press, 2006). See also the chapters in Pamela H. Smith and Benjamin Schmidt (eds), *Making Knowledge in Early Modern Europe* (Chicago, IL: University of Chicago Press, 2007).

11 Pioneering works include Jennifer Stine, *Opening Closets: The Discovery of Household Medicine in Early Modern England* (PhD diss., Stanford University, 1996), pp. 176–215; Lynette Hunter, 'Women and Domestic Medicine: Lady Experimenters, 1570–1620', in Lynette Hunter and Sarah Hutton (eds), *Women, Science and Medicine, 1500–1700: Mothers and Sisters of the Royal Society* (Thrupp: Sutton Publishing, 1997), pp. 89–107; Laura L. Knoppers, 'Opening the Queen's Closet: Henrietta Maria, Elizabeth Cromwell, and the Politics of Cookery', *Renaissance Quarterly*, 60 (2007): pp. 464–99; Sara Pennell, 'Perfecting Practice? Women, Manuscript Recipes and Knowledge in Early Modern England', in Victoria Burke and Jonathan Gibson (eds), *Early Modern Women's Manuscript Writing: Selected Papers from the Trinity/Trent Colloquium* (Aldershot: Ashgate, 2004), pp. 237–358; Catherine Field, '"Many hands hands": Writing the Self in Early Modern Women's Recipe Books', in Michelle M. Dowd and Julie A. Eckerle (eds) *Genre and Women's Life Writing in Early Modern England* (Aldershot: Ashgate, 2007), pp. 49–63; Elaine Leong and Sara Pennell, 'Recipe Collections and the Currency of Medical Knowledge in the Early Modern "Medical Marketplace"' in Mark S. Jenner and Patrick Wallis (eds), *Medicine and the Market in England and its Colonies c. 1450–c. 1850* (Basingstoke: Palgrave Macmillan, 2007), pp. 133–52; Elaine Leong, 'Making Medicines in the Early Modern Household', *Bulletin of the History of Medicine*, 82/1 (2008):

Finally, Jo Wheeler and Allison Kavey have both recently addressed the topic of secrets head-on.[12] Not all of these authors characterize themselves as studying 'secrets', nor should they all be seen as derivative of Eamon's work. Yet the result of this collective body of scholarship has been a normalization of technical, craft and empirical knowledge as a topic of study and a recognition that curiosity about this kind of practical knowledge crossed into a huge variety of disciplines, interests and social circles.

As a consequence of this large array of diverse scholarship, there is no longer any question of the importance of secrets to early modern medicine and science. As Eamon describes in Chapter 1 of this volume, he received blank stares in response to his first paper on books of secrets at a professional conference in 1982. In 2008, in contrast, the University of Cambridge's Centre for Research on the Arts, Social Sciences and Humanities hosted an entire symposium devoted to secrets' role in early modern scientific and medical knowledge, the basis of this volume. The range of topics presented at the conference and in these chapters make it clear just how far the field has come. Past studies, for example, have tended to focus on the sixteenth century. The chapters in this volume, in contrast, cover a long chronological period stretching from late medieval Spain to Enlightenment France, highlighting both continuities and changes in the way secrets and secrecy functioned in the making of early modern medical and scientific knowledge.

The authors in this volume expand not only the chronology, but also the source base for the study of secrets. The sixteenth century has received attention because it represented the golden period for the print genre 'books of secrets', identified by Ferguson and later defined and explicated by Eamon and, most recently, Kavey. The majority of editions and translations of Alessio's *I secreti* were produced between 1550–1600, and other professors of secrets, such as Leonardo Fioravanti, Johann Jacob Wecker and Hugh Platt, also offered their work to readers in the latter half of the sixteenth century. Printed books of secrets are generally associated with these (and other) learned or professional men, who operated, for the most part, within established institutions. However, print represented only one medium in which secrets were produced and exchanged, and there was an equally vibrant engagement with secret knowledge in manuscript sources. Broadening the focus beyond print allows our contributors to vastly expand this purview, adding new actors to the traditional cast of characters and demonstrating the depth and longevity of interest in secrets.

Secrets and Knowledge in Medicine and Science, 1500–1800 illustrates the centrality of secrets in various arenas of early modern European knowledge

---

pp. 145–68; Alisha Rankin, 'Duchess, Heal Thyself: Elisabeth of Rochlitz and the Patient's Perspective in Early Modern Germany', *Bulletin of the History of Medicine*, 82/1 (2008): pp. 109–44.

12 Jo Wheeler, *Renaissance Secrets, Recipes, and Formulas* (London: V&A Publishing, 2009); Allison Kavey, *Books of Secrets: Natural Philosophy in England, 1550–1660* (Urbana and Chicago, IL: University of Illinois Press, 2007).

making by bringing together several strands of research, including the history of medicine and health, artisanal and craft history, book history and the history of commerce. The volume has three main goals. First, it demonstrates the extent to which the research on secrets has expanded beyond 'books of secrets'. While early modern print culture is still a prominent feature in these chapters, our contributors scoured the archives and examined a range of manuscript sources from letter collections to medical casebooks to household recipe books. Second, it highlights the diversity of what counted as a secret, whether this meant directives to make everyday foodstuffs and medicines or a set of alchemical instructions or technical trade know-how. Finally, it demonstrates that national and historical contexts must be taken into account in creating a narrative of the historical 'secret', for approaches to the term in Renaissance Venice differed substantially to those in Restoration England. Read as a whole, these chapters demonstrate the variety of contexts in which secrets – as little bits of information – played a central role in producing and sharing knowledge in early modern Europe.

The ten chapters in this volume provide a number of different frameworks to explore secrets and knowledge in early modern Europe. In the first part, 'Defining Secrets', William Eamon's 'How to Read a Book of Secrets' encourages us to ponder how these texts might have been received by contemporary readers and how we, as historians, might usefully employ them in schemas of research. Pamela H. Smith's 'What is a Secret?' pushes us to analyze recipes for traces of 'vernacular science', the theoretical frameworks and principles used by craftsmen to understand matter and nature. Her chapter underscores the key role that craft knowledge played in the interest in secrets.

'Secrecy and Openness', the second part, presents three studies of knowledge networks in early modern England. Ayesha Mukherjee examines 'The Secrets of Sir Hugh Platt', an Elizabethan author and inventor famous for his printed books of secrets. Mukherjee's careful readings of Platt's writings, both in manuscript and print, situate his experimental pursuits within contemporary anxieties of dearth and economic problems. In 'Robert Boyle and Secrecy', Michael Hunter presents a detailed investigation of Boyle's shifting approach to secrets and secrecy, depending on his state of involvement with experimental sciences. Boyle's case exemplifies the optimism about demystifying the secrets of nature as well as the drive to keep the most valuable secrets secret, even for those who professed an ideal of openness. Similar concerns about when to reveal and when to disclose secrets also underlie Michelle DiMeo's chapter, 'Openness vs. Secrecy in the Hartlib Circle'. Focusing on a group well known for their pleas to make available all useful knowledge, DiMeo explores instances in which secrecy was nevertheless viewed as necessary. Her chapter underscores the different contexts, social and intellectual, that delineated both the need for openness and for secrecy.

In the third part, 'Illicit Secrets', we encounter two women who, working in very different environments, proffered an array of secrets. In 'Anna Zieglerin's Alchemical Revelations', Tara Nummedal paints a colourful account of how

Anna Zieglerin used controlled disclosure of secret alchemical knowledge to establish her position within the court. Despite Zieglerin's grisly end, her complex interweaving of alchemical secrets with craft secrets, secrets of state, secrets of nature and divine revelation serves as a reminder of the potency of secret knowledge in early modern Europe. Tessa Storey's 'Face Waters, Oils, Love Magic and Poison' brings to our attention the figure of Maddalena, a woman from seventeenth-century Rome who wore many hats: seamstress, landlady, procuress and the go-to person for love potions and cures. Maddalena's trial records provide a rare glimpse of how 'secrets', as bits of know-how and as made up remedies and potions, circulated and functioned within an urban Italian setting. Her detailed case study not only further emphasizes the commercial aspects of making and keeping secrets but also highlights the multiple layers of meanings imposed upon secrets and knowledge.

The final part in our volume, 'Secrets and Health', contains chapters that highlight three hitherto understudied areas of early modern medicine and healthcare activities. Montserrat Cabré's work on Juan Vallés' unpublished *Regalo y policía de la vida humana* (c. 1563), 'Keeping Beauty Secrets in Early Modern Iberia', not only draws attention to the importance of cosmetic secrets but also locates the making of medical knowledge within the world of book production in the early days of print. Sandra Cavallo focuses upon 'Secrets to Healthy Living' and reminds us of the central place regimens and health maintenance held in early modern European minds. Finally, Lisa Wynne Smith's chapter, 'Secrets of Place', uses the physician Vivant-Augustin Ganiare's records of local weather and disease patterns to place medicine and secrets within Enlightenment religious and intellectual discourses. Her analysis of Ganiare's medical casebooks, a text genre not traditionally considered for the study of secrets, demonstrates how the physician was able to combine his knowledge of scientific methodology, his religious convictions and patience and perseverance to learn the secrets of nature.

Defining Secrets

What are secrets? How and why did they figure into early modern science and medicine? These questions are central to our volume, yet pinning down a simple answer is next to impossible. 'Secrets' and secret knowledge can be found in many overlapping spheres of early modern medicine, natural philosophy and the technical arts, and specific ways in which the terms were invoked depended highly on cultural context. Secrets and secrecy had long played a central role in the transmission of esoteric and occult knowledge, particularly in the alchemical tradition. The pseudo-Aristotelian *Secretum secretorum*, which purported to reveal the hidden, esoteric knowledge of the great philosopher, was one of the most popular texts of the Middle Ages. This type of secret knowledge was closely connected to the idea of the mystery of God's creation, revealed only to a chosen few.[13] Similarly, secrets were associated with wondrous events and objects, whether these occurred in nature or were created through human manipulation of nature.[14]

Yet alongside this invocation of hidden and mysterious knowledge, the word 'secret' could also refer more specifically to a set of procedures known only to a select group of initiated individuals – in other words, craft or trade secrets. Recent scholarship has emphasized the importance of trade secrets shared among artisans, medical practitioners, early modern technical operators and engineers.[15] This kind of secret was more about technical know-how, or 'how to', than hidden knowledge (although the two concepts were by no means mutually exclusive). Technical, craft-based secrets were key to transmitting knowledge about medicine and science, as they provided directions for completing specific processes. They proved so popular in the sixteenth century that Eamon deemed it the 'Age of How-to'.[16] In most cases, secrets of this nature took the form of a recipe.

Recipes, typically sets of short matter-of-fact instructions, constituted one of the main text forms through which early modern practical knowledge circulated. Today recipes tend to have close associations with kitchens and foodstuffs, but in early modern Europe they were used whenever practical know-how needed to be transmitted, including topics as diverse as pharmacy, painting, metallurgy, juggling, smithing, fireworks and gunnery. The practical knowledge inherent in recipes did not preclude occult or esoteric knowledge, as demonstrated by Nummedal's study of Anna Zieglerin. However, the possession of a recipe allowed the holder the promise of a potentially valuable result. In *Science and the Secrets of Nature*, Eamon called a recipe 'a prescription for an experiment; a "trying-out"'.[17] Although he modifies that view in this volume, his essential point about the connection between the recipe and the action it prescribes remains relevant. Perhaps for this reason, collecting recipes became a faddishly popular enterprise all across early modern Europe, from the princely courts to humanist libraries to artisanal circles to country house kitchens.

Hence secrets and recipes were both key to the production and transfer of natural knowledge in early modern Europe, with circulation numbers in the hundreds of thousands. But were all recipes secrets? What made a recipe a 'secret'

---

13 Eamon, *Science*, pp. 45–53.

14 Lorraine Daston and Katharine Park, *Wonders and the Order of Nature 1150–1750* (New York, NY: Zone Books, 2001).

15 See, for example, Long, *Openness, Secrecy, Authorship*. See also the chapters in the following edited collections: Lissa Roberts, Simon Schaffer and Peter Dear (eds), *The Mindful Hand: Inquiry and Invention from the Late Renaissance to Early Industrialization* (Amsterdam: Koninklijke Nederlandse Akademie van Wetenschappen, 2007) and Smith and Schmidt, *Making Knowledge*.

16 Eamon, *Science*, pp. 126–33.

17 Ibid., p. 131.

and not just a list of ingredients with accompanying instructions? In some contexts, there was in fact little difference. In Renaissance Italy, for example, the words 'secret' and 'recipe' were virtually synonymous, and 'secret' could be used to mean 'recipe' with no connotation of secrecy. Far more often, however, the word 'secret' was invoked with specific intent. In the stylized preface of Piemontese's *I secreti*, Alessio claimed he spent most of his life collecting secrets for himself alone, until his reticence to disclose them led to the death of a man who could have been saved by them. By describing this dramatic event, Alessio depicted his secrets as formerly esoteric knowledge that had great value, finally revealed to you, the reader. His secrets consisted entirely of recipes, but they were recipes that disclosed previously hidden (and valuable) knowledge.

While the author of Alessio's *I secreti* (Ruscelli) invoked secrecy as a device to entice the reader, Anna of Saxony used it to create bonds with other noblewomen. As we have seen, Anna claimed that she had shared her 'most secret arts' (*geheimbste kunste*) with Duchess Anna of Bavaria. In so doing, she created the sense of a shared community of privileged knowledge. Building communities of knowers was one of the most crucial functions of secrets. Within a given circle of shared knowledge, whether an artisanal guild, a family, the Royal Society or a group of gentlewomen, secrecy was stressed in relationship to outsiders, while openness was usually encouraged in exchanges among its members. This contrast between secrecy and openness, a key theme in both Hunter's and DiMeo's chapters, fostered the sense of a community of knowledge-builders. While early modern scientific circles often used the trope of openness to contrast themselves with esoteric Hermetic cliques of the Middle Ages, in fact the idea that knowledge should be extended only to a chosen few remained alive and well.

In many cases deliberate steps were taken to conceal secrets from unwelcome eyes. Encryption, a practice common in the transmission of alchemical knowledge, was a particularly common device. As Hunter demonstrates in this volume, even as revered a scientific figure as Robert Boyle used standard encryption methods, including word substitution and dispersal, to ensure secrecy in his workdiaries. Such methods were also employed by lesser-known figures in their dealings with more mundane information. In mid-seventeenth-century England, Sir Peter Temple created a notebook of medical recipes for his daughter Eleanor. Unusually, Temple encrypted his remedy collection by writing keywords of certain recipes (authors' names and ingredients to be used) in a simple cipher. The reason for this, he explained to Eleanor, was threefold: firstly, to 'conceale it if it come in to an Enemys hand', secondly 'because Even Friends doe dispise w^{t} they know' and finally 'to keepe ignorant persons from applying y^{m} (too often) to the Patients prejudice'.[18] The knowledge contained in his book, he suggested, was too sensitive to be shared with just anybody. Intriguingly, Temple encrypted not only recipes from family and friends, but also information he copied from printed books. He explained to Eleanor that 'the receits markt *GM* are *printed* and are *Mr Markhams*

18 British Library, MS Stowe 1077, fol. 10r.

and yet I estemme them not the lesse', a reference to Gervase Markham, author of *The English Housewife* (1615) and other popular Tudor household guides.[19] As Markham's books were bestsellers in Temple's day, the information presented can hardly be classified as top secret or any sort of secret. Instead, the secrecy surrounding these recipes was generated and imposed by the reader-turned-author Temple.

Why would a little-known English gentleman take the time to encrypt recipes, particularly remedies widely available in print? While we can never know for certain, his cipher underscores several key points in understanding secrets as used throughout early modern Europe. Most obviously, despite the wide availability of Markham's books, there was still value in picking out remedies deemed particularly useful. Like Alessio's secrets, the rendering of this knowledge as concealed and 'secret' increased the perceived value and authority of the remedies. In addition, it created a community of shared knowledge of the sort discussed above. Undoubtedly, when Temple bequeathed his recipe book to his daughter, he accompanied it with the key to his cipher, thus inviting Eleanor and her family into his circle of privileged knowers. The steps he took to keep his most favoured remedies within the family were indicative of a widespread drive, taken from prominent groups such as the Royal Society and the Hartlib circle on down to much smaller and less notorious communities, to protect valuable knowledge. Steven Shapin and Simon Schaffer have detailed ways in which virtual communities shared information; just as important were the ways in which they protected privileged information.[20] As the Duchess Anna Maria of Württemberg put it succinctly, 'what every man has is no longer an art'.[21]

In a variety of ways, then, authors consciously designated knowledge they deemed particularly noteworthy as 'secret'. The cloak of secrecy could be drawn for a range of reasons – enticing new readerships, creating virtual intellectual communities and making a recipe appear particularly noteworthy. As the chapters in this volume emphasize, the relationship between secrets and recipes was complex, ever changing and highly dependent on context. The majority of our authors take secrets to be recipes endowed with special value, although the reasons for assigning that value differ in each chapter. Nonetheless, there are also contrasting interpretations. Cavallo, working on Renaissance Italian physicians' texts and craft manuals, sees a nearly synonymous relationship between 'secret' and 'recipe', while Storey notes a remaining connection between the word 'secret' and the occult in her sources from the lower classes of seventeenth-century Rome. Valuable particular knowledge of the kind possessed by Anna

---

[19] Ibid., encoded words are italicized.

[20] Steven Shapin and Simon Scaffer, *Leviathan and the Air Pump: Hobbes, Boyle and the Experimental Life* (Princeton, NJ and Chichester: Princeton University Press, 1985).

[21] 'was ider man hat ist es kan kunst mer'. Sächsische Hauptstaatsarchiv, Dresden, Geheimer Rat (Geheimes Archiv), Loc. 8529/2, fol. 41v.

Zieglerin (Nummedal) or Vivant-Augustin Ganiare (Wynne Smith) is another, slightly different use of the term.

Where does that leave us? What was a secret? As we can see, no one definition will ever satisfy, given how much it depended on contemporary authors, readers and users. For it was the author's/reader's/user's decision to identify particular types of knowledge as 'secret' that elevated a mere recipe to a secret – or, in some cases, made virtually no distinction between the two concepts. Consequently, rather than attempting to draw strict delineations between secrets and recipes, it is perhaps more useful to further explore the different contexts (intellectual, economical, social, political) which encouraged early modern writers to identify particular knowledge as such.

Secrets in Medicine, Science and Commerce

As the title of this book suggests, medical and scientific secrets and secret knowledge were intimately bound up in early modern commerce, trade and markets. In recent years, historians such as Pamela H. Smith, Paula Findlen, Harold Cook and Antonio Barrera-Osorio have demonstrated the important relationship between commerce and the production of natural knowledge.[22] The flourishing of trade between Europe and the New World brought with it encounters with unknown natural and artificial objects and lead to a rapid gathering of new natural knowledge.[23] At the same time, trade with the Ottoman Empire and southeast Asia continued to thrive, and Jesuit and trade expeditions to China and Japan presented an opportunity for new information on these once-mysterious territories.[24] The European enthusiasm for new botanical and

22 Pamela H. Smith and Paula Findlen (eds), *Merchants and Marvels: Commerce, Science and Arts in Early Modern Europe* (New York, NY and London: Routledge, 2002); Harold Cook, *Matters of Exchange: Commerce, Medicine and Science in the Dutch Golden Age* (New Haven, CT and London: Yale University Press, 2007); and Barrera-Osorio, *Experiencing Nature*.

23 Antonio Barrera-Osorio, 'Local Herbs, Global Medicines: Commerce, Knowledge, and Commodities in Spanish America', in Smith and Findlen, *Merchants and Marvels*, pp. 163–81; Barrerra-Osorio, *Experiencing Nature*, esp. pp. 81–101; Steven J. Harris, 'Networks of Travel, Correspondence and Exchange', in Katharine Park and Lorraine Daston (eds), *The Cambridge History of Science, vol. 3: Early Modern Science* (Cambridge: Cambridge University Press, 2006), pp. 341–60; Cook, *Matters of Exchange*; Daniela Bleichmar, 'Books, Bodies and Fields: Sixteenth-Century Transatlantic Encounters with New World Materia Medica', in Londa Schiebinger and Claudia Swan (eds), *Colonial Botany: Science, Commerce, and Politics in the Early Modern World* (Philadelphia, PA: University of Pennsylvania Press, 2005), pp. 83–99.

24 Cook, *Matters of Exchange*, Ch. 9; Klaus A. Vogel, 'European Expansion and Self-Definition', trans. Alisha Rankin, in Park and Daston, *Cambridge History of Science*, pp. 821–40.

zoological specimens and other forms of *naturalia* is not only evident in the detailed descriptions in print and in manuscript but also in the widespread collecting practices and the numerous cabinets of curiosities created during this period. Unsurprisingly, this drive to collect *naturalia* fit in very closely with the impetus to find and propagate the secrets of nature.[25] For some readers and compilers of secrets, the act of collecting these tidbits of practical or technical information was part of individual schemes to understand the natural world around them.

Within medicine, secrets and secret knowledge occupied an important role in delineating and driving the development of new knowledge, techniques and medical economies. Proprietary knowledge of particular drugs, cures, procedures and instruments allowed a wide range of practitioners, from apothecaries to midwives to mountebanks and charlatans, to stand out within the bustling medical marketplace.[26] As Mukherjee and Storey demonstrate in this volume, secrets were often for sale as physical objects. Historians David Gentilcore, Alison Klairmont-Lingo and Roy Porter have described early modern piazzas and squares bustling with charismatic travelling medical practitioners who used colourful shows to sell their nostrums.[27] Aside from itinerant mountebanks, charlatans and 'quacks', licensed medical practitioners such as physicians and apothecaries also began to issue proprietary medicines. One such drug, Daffy's Elixir Salutatis, was an instant success. Produced in large quantities, it was shipped and sold in not only in England but Scotland, Ireland, France, Holland and New England.[28] The formula or recipe for the elixir was closely guarded by the Daffy family but, inevitably, the secret got out and various versions of the recipe circulated in both manuscript and print in the period. Antony Daffy and his family's financial success, however, was based on their claim that their own recipe, unadulterated by circulation and copyists' error, was the 'secret' and the superior version. A secret could thus be a physical object (a remedy) as well as the knowledge required to make that object.

---

[25] Paula Findlen, *Possessing Nature: Museums, Collecting, and Scientific Culture in Early Modern Italy* (Berkeley, CA: University of California Press, 1994); and chapters in Smith and Findlen, *Merchants and Marvels*.

[26] See especially Jenner and Wallis, *Medicine and the Market*.

[27] David Gentilcore, *Medical Charlatanism in Early Modern Italy* (Oxford: Oxford University Press, 2006); Alison Klairmont Lingo, 'Empirics and Charlatans in Early Modern France: The Genesis of the Classification of the "Other" in Medical Practice', *Journal of Social History*, 19/4 (1986): pp. 583–603; Roy Porter, *Health for Sale: Quackery in England 1660–1850* (Manchester and New York, NY: Manchester University Press, 1989) and William Eamon, 'Markets, Piazzas and Villages', in Park and Daston, *Cambridge History of Science*, pp. 206–23.

[28] David B. Haycock and Patrick Wallis, 'Quackery and Commerce in Seventeenth-century London: The Proprietary Medicine Business of Antony Daffy', *Medical History*, Supplement no. 25 (2005), p. 16.

Many of these healers took advantage of print – still a relatively new technology – to advertise or manage this kind of proprietary secret. Italian charlatans were rarely seen without their printed single-sheet handbills or flyers, a medium also used by surgeons, apothecaries and theriac sellers. Broadsheets, pamphlets, chapbooks and, by the eighteenth century, newspapers were also common venues for advertising secrets in Italy and elsewhere.[29] Mukherjee has discovered that Hugh Platt's successful book, *The Jewell House of Art and Nature* (1594), was a reference to the eponymous shop ('Jewel House') where he sold his secrets. While it is unclear to what extent Platt intended to advertise his shop, other publications were less subtle. We have already seen that the 'expert operator' Edward Fountain used a printed book of secrets to promote his London-based dental practice in the mid-seventeenth century. In fact, as Adrian John has noted, there was an extremely close alliance between printers and medical practitioners in early modern London: healers from physicians to 'expert operators' relied on booksellers to advertise new drugs, and some printers even prepared and sold the medicines in their shops.[30] Similarly, the German lawyer-turned-empiric Georg am Wald used a book printed in 1591 to publicize the merits of his cure-all, the 'Panacea Amwaldina', and to inform the reader where one could buy it – but did not reveal the recipe, which remained a secret.[31] In other cases, the publication of secrets was meant to exert control over a medicament already in broad circulation. In his chapter in this volume, Eamon suggests that Leonardo Fioravanti used print to create de facto patents for his medicines as a means to manage the widespread sale of his medical formulas. Only by following his recipe to the letter, Fioravanti insisted, could the reader be assured of success.

While self-promotion might have been the driving force in the relationship between medicine, secrets and commerce, some endeavours also made new practical and technological knowledge available to readers for public and common good. Platt, for example, was greatly concerned with the agricultural and economic difficulties of 1590s England. The problem of dearth, in particular, motivated his engagement with natural knowledge and the publication/sale of his secrets. Similarly, physicians' secrets for healthy living, as described by Cavallo and Lisa Smith, were purportedly shared for the good of the patient and/or young physicians, with the goal of preserving the health of the individual. The physician

---

[29] Gentilcore, *Medical Charlatanism*, pp. 337–69.

[30] Adrian Johns, *Piracy: The Intellectual Property Wars from Gutenberg to Gates* (Chicago, IL: Chicago University Press, 2009), pp. 83–108 (p. 84).

[31] Georg am Wald, *Kurtzer Bericht, wie und was gestalt der Panacea Amwaldina als eine einige Medicin ... anzuwenden sey* (Frankfurt am Main, 1591). See also Alisha Rankin, 'Empirics, Physicians and Wonder Drugs in Early Modern Germany: The Case of the *Panacea Amwaldina*', *Early Science and Medicine*, 14/6 (2009): pp. 680–710; Wolf-Dieter Müller Jahncke, 'Georg am Wald (1554–1616)', in Joachim Telle (ed.), *Analecta Paracelsica: Studien zum Nachleben Theophrast von Hohenheims im deutschen Kulturgebiet der frühen Neuzeit* (Stuttgart: Franz Steiner Verlag, 1994), pp. 213–304.

Ganiare, as Smith describes, specifically saw his health advice as a form of medical charity. Yet even in projects intended for wider benefit, there was a complex relationship between business interests and a genuine desire for collaborative problem-solving or the greater good. Platt's *Jewel House*, after all, sold many of the objects described in his publications, and Cavallo's physicians sold their books of regimen. Even Ganiare's casebooks, which the author bequeathed to the town of Beaune as a charitable gift, were later sold (and charity in itself was an important form of commerce). As units of proprietary knowledge, secrets were – sometimes explicitly, sometimes less so – inexorably intertwined with commercial considerations.

New Forms, New Spaces

Print was a particularly popular venue for advertising and communicating secrets, but it was not the only medium used in the circulation of secret knowledge. As the chapters in this volume show, manuscript sources were crucial for the production and communication of such information. The introduction of print in the second half of the fifteenth century and the subsequent boom of the print trade brought with it possibilities for commercial opportunities and broader audiences. Yet despite these advantages of scale, manuscripts continued to be preferred for the person-to-person benefits that print lacked. Fioravanti, as Eamon has recently described, ran a successful mail-order practice in which patients would recount their ailments in letters and await his advice and 'important secret remedies'.[32] The German empiric Georg am Wald received a deluge of personal communication following the publication of his work on the 'Panacea Amwaldina' and appended 70 pages of the letters to the book's third edition, thereby taking advantage simultaneously of the personal nature of manuscript culture and the broad dissemination of print.[33] The printed books of secrets themselves were based largely on manuscript correspondence and personal and craft notebooks, and many authors, such as the venerable Robert Boyle, preferred not to publish certain bits of secret knowledge (see Hunter).

The chapters in this volume highlight the sheer variety of manuscript sources in which secrets were written down, exchanged, discussed and passed on. Hunter's contribution draws on Boyle's workdiaries and correspondence to demonstrate the

---

[32] William Eamon, *The Professor of Secrets: Mystery, Medicine and Alchemy in Renaissance Italy* (Washington, DC: National Geographic, 2010), p. 220.

[33] Georg am Wald, *Kurtzer und zum andernmal gemehrter Bericht von der Panacea Amwaldina* (Ursel, 1594). Andreas Libavius took such exception to the inclusion of his letter of mere curiosity amongst the letters of praise that he penned a Latin treatise decrying am Wald, which he subsequently published in German. Bruce Moran, *Andreas Libavius and the Transformation of Alchemy: Separating Chemical Cultures with Polemical Fire* (Sagamore Beach, MA: Science History Publications, 2007), pp. 125–32.

contrast between his published push towards openness and his private inclinations towards secrecy. Letters and correspondence play a similarly important role in Michelle DiMeo's analysis of openness and secrecy within the Hartlib circle and Tara Nummedal's examination of the German alchemist Anna Zieglerin. Another key source, the household recipe book, has received considerable attention in the past few years and features in several of the chapters in this volume, including Cabré's detailed study of a fifteenth-century Iberian collection of beauty recipes, Cavallo's focus on preservative remedies in late Renaissance Italy and Mukherjee's exploration of Platt's numerous notebooks. Manuscript books of craft secrets such as Cenino Cennini's *Il libro dell'arte* are utilized by Pamela H. Smith to argue for a 'vernacular philosophy', a set of principles craftsmen used to orientate themselves within their understanding and manipulation of nature, while Lisa Wynne Smith draws on the medical casebooks and personal accounts of the physician Ganiare to explore the point of view of an Enlightenment physician. Finally, Nummedal and Tessa Storey pore over court trial transcripts to reconstruct two very different cases of women accused of peddling medical and alchemical secrets.

The intersection of manuscript, print and oral cultures is a central feature in a number of the chapters, further emphasizing the extent to which secrets and secret knowledge were transmitted in all three media. Each form brought with it different intended (and actual) audiences, circulation consequences and cultural associations. Person-to-person exchanges of tacit knowledge were one of the most important means of communicating secrets, and Pamela H. Smith's chapter investigates the relationship between craft practice and practical guides to get at this question of trade knowledge inscribed in writing. The significance of writing was not lost on early modern individuals. When the Roman seamstress Maddalena's neighbours were called to testify against her, Storey notes, they were keen on stressing that Maddalena knew 'great written secrets' (*gran secreti scritti*). The phrase emphasizes the contemporary view that boundaries existed between oral and written knowledge and that writing had a special power.

At the other end of the spectrum, secrets publicized in print were often viewed as less valuable or proprietary than those confined to manuscripts. For example, Regina Zangmeister (d. 1597), wife of an Augsburg merchant, sent a book of handwritten medicinal recipes to a Bavarian countess with the assurance that they had not been 'taken or copied from printed books, of which everyone can obtain an exemplar for money'. Instead, they had originated from 'eminent, honest people of good name … written in their own hands'.[34] Similarly, the compiler or scribe of a late seventeenth-century English household book of recipes boasted that none of the recipes contained within 'have been ever published in print but in Parcells Transmitted in writing to diverse families'.[35] In such cases, the shift

[34] The manuscript was addressed to Countess Claudia of Oettingen-Oettingen. Universitätsbibliothek Heidelberg, Codex Palatinum germanicum 227, fols 4ar-5ar.

[35] Bethesda, MD, National Library of Medicine, F 98, fol. 1r.

between orality/writing and manuscript/print was perceived as a step imbued with meaning.

The transition from manuscript to print was not direct or straightforward for early modern book producers or consumers.[36] A number of the textual genres within which we find secrets and secret knowledge, such as recipe collections, regimen guides and medical casebooks, were, of course, longstanding manuscript traditions. When adapting these texts for print, book producers made important decisions in regards to textual content, organization and presentation. Montserrat Cabré's examination of Juan Vallés' *Regalo y policía de la vida humana* – a manuscript intended for print but never printed – demonstrates that the reordering of a knowledge schema could stem from both natural philosophical endeavours and practical publishing needs. Investigations of the intersection between script and print also lie at the centre of studies into book use and consumption. In recent years, historians of reading have demonstrated the fruitfulness of examining traces of reading both in the form of marginal annotations or books of reading notes.[37] In this volume, this kind of exploration can be seen in Mukherjee's study of Platt's library, his notebooks and his published books where she teases out a reader's engagement with both manuscript and printed texts and his subsequent output as an author. As with most individuals introduced here, the boundaries between reader, experimenter, compiler, author and user are malleable and renegotiated with each instance of sharing secret knowledge.

A number of these transactions took place within spaces not traditionally recognized as sites of knowledge making. The systematic annotator Platt and the readers of the highly marked-up copies of Alessio's *I secreti* all made the most of the wide margins in early modern printed books. For these readers, margins became the space for formulating new knowledge. In the past 20 years, the 'spaces' of early modern science have been continually expanded, with princely courts and printing houses now widely accepted as places in which knowledge

---

36 Adrian Johns, *The Nature of the Book: Print and Knowledge in the Making* (Chicago, IL and London: University of Chicago Press, 1998) and David McKitterick, *Print, Manuscript and the Search for Order 1450–1830* (Cambridge: Cambridge University Press, 2003).

37 See for example, Roger Stoddard, *Marks in Books, Illustrated and Explained* (Cambridge, MA: Houghton Library, Harvard University, 1985); Lisa Jardine and Anthony Grafton, '"Studied for Action": How Gabriel Harvey Read His Livy', *Past and Present*, 129 (1990): pp. 30–78; Anthony Grafton, *Commerce with Classics: Ancient Books and Renaissance Readers* (Ann Arbor, MI: University of Michigan Press, 1997); Heather Jackson, *Marginalia: Readers Writing in Books* (New Haven, CT: Yale University Press, 2001); Heidi Brayman Hackel, *Reading Material in Early Modern England: Print, Gender and Literacy* (Cambridge: Cambridge University Press, 2005); Bradin Cormack and Carla Mazzio, *Book Use, Book Theory 1500–1700* (Chicago, IL: University of Chicago Library, 2005) and William Sherman, *Used Books: Making Readers in Renaissance England* (Philadelphia, PA: University of Pennsylvania Press, 2008).

was created, alongside the traditional settings of universities and academies.[38] Most recently, this expansion has begun to include household studies, kitchens and stillrooms.[39] Expanding the sites of science is especially fruitful in the study of secrets. The past concentration upon printed books of secrets has led to a focus on formal institutional spaces as locations for the creation of and circulation of secrets – for example, Eamon writes of Ruscelli's *academia segreta*, within which 27 members dedicated their days to experimental research. Yet sources such as the domestic manuscript notebooks of secrets and recipes, annotated books of secrets and personal writings suggest the importance of the early modern household in transferring and producing secrets. The chapters in this volume demonstrate that, for personalities as diverse as Robert Boyle (Hunter), the seamstress Maddalena (Storey), Hugh Platt (Mukherjee), or the physician Ganiare (Wynne Smith), household 'laboratories' and private studies played a central role in propagating early modern secrets.

New Authors

By opening up the documentation of early modern secrets to include manuscript sources and by acknowledging the household as an important site in which they were produced, we add a crucial demographic that is nearly invisible in printed sources: female authors and participants. With one notable exception, all of the books of secrets studied in Eamon's *Science and the Secrets of Nature* were ascribed to men. That exception, a very popular book titled *I secreti de la Signora Isabella Cortese* (Venice, 1561), remains something of a mystery, and it is still unclear whether a woman named Isabella Cortese was the actual author of the work – or even a real person.[40] Even if Cortese's work is taken at face value, there is no doubt that the quintessential 'professor of secrets' is a man. Women do not completely lack representation in printed books of secrets: reading between the lines, one can see evidence of women's involvement in many of these works. Alessio Piemontese, for example, claimed to have gotten his secrets from 'poor women' alongside artisans and men of great learning. The fact that Isabella Cortese's book went through seven editions by 1600, moreover, highlights the

---

38 Bruce Moran (ed.), *Patronage and Institutions: Science, Technology, and Medicine at the European Court, 1500–1750* (Rochester, NY: Boydell Press. 1991).

39 Alix Cooper, 'Homes and Households', in Daston and Park, *Cambridge History of Science*, pp. 224–37; Hunter and Hutton, *Women, Science and Medicine*; Deborah E. Harkness, 'Managing an Experimental Household: The Dees of Mortlake and the Practice of Natural Philosophy', *Isis*, 88/2 (1997): pp. 247–62; Rankin, 'Becoming an Expert Practitioner: Court Experimentalism and the Medical Skills of Anna of Saxony (1532–85)', *Isis*, 98/1 (2007): 23–53.

40 Isabella Cortese, *I secreti dela signora Isabella Cortese* (Venice: Giovanni Bariletto, 1565).

attraction of secrets revealed by a woman, whether or not Cortese was the true author. In most printed books of secrets, however, women hover just beneath the surface.

Manuscript sources, in contrast, add a rich diversity of female authors and compilers of secrets. Many, like Anna of Saxony, were gentlewomen and had large court distilleries at their disposal. Anna, for example, had a distilling house with 12 ovens, and she engaged in avid exchange of 'secret arts' with aristocratic women and men, as well as some physicians. Lynette Hunter has highlighted a similar profusion of 'lady experimenters' in England, most prominently Lady Katharine Ranelagh, who kept up a vibrant knowledge exchange with her brother, Robert Boyle.[41] As DiMeo demonstrates in this volume, Ranelagh was also an avid participant in the Hartlib circle and exchanged secrets with its other members. Interest in secrets was by no means limited to the upper classes. Other women, like Maddalena of Rome, hailed from much more modest means yet still mastered 'great written secrets'. Beyond the scope of printed sources lay a world in which women were frequent and widely accepted sources of secrets and knowledge.

At the same time, it should not escape our attention that the two studies of female purveyors of secrets in our volume can be found in the section titled 'Illicit Secrets'. Nummedal and Storey both present cases of women who peddled secrets, fell afoul of the law, and were put on trial. Their cases are suggestive of the diatribes against 'women's secrets' in sources dating back to the Middle Ages. As Katharine Park notes, the term 'secrets of women' had a twofold meaning: firstly, knowledge of the female reproductive system, which was internal and thus a mystery, and secondly, therapeutic knowledge possessed by certain women that was 'orally transmitted, experience-based, concrete, and bodily oriented'.[42] This latter kind of secret, while certainly not the sole province of women, presented a convenient contrast with scholastic natural philosophy, which was based on the texts of ancient Greek and Arabic sources and was associated with learned men. From the Middle Ages through at least the seventeenth century, women – and particularly old women or *vetule* – feature prominently in physicians' invectives against empirical practitioners.

The cases of Anna Zieglerin and the seamstress Maddalena would seem to fit into this long tradition of denigrating women's secrets as a lower form of knowledge. Zieglerin, whose secrets partially involved her body's reproductive capabilities, also plays into the view of women's secret knowledge as closely connected to generation and reproduction, a connection studied convincingly by

---

41 Lynette Hunter, 'Women and Domestic Medicine: Lady Experimenters 1570–1620' and 'Sisters of the Royal Society: The Circle of Katherine Jones, Lady Ranelagh', in Hunter and Hutton, *Women, Science and Medicine*, pp. 89–107 and 178–97.

42 Katharine Park, *Secrets of Women: Gender, Generation and the Origins of Human Dissection* (Cambridge, MA: MIT Press, 2007), pp. 82–3, (p. 82).

Park, Susan Broomhall and Monica Green.[43] Zieglerin and Maddalena remind us that the freedom with which women like Electress Anna of Saxony or Lady Katharine Ranelagh pursued their interests in secrets can by no means be taken as the norm. Yet it would be far too simple to read these cases merely as evidence of a continued suspicion of the 'secrets of women'. In Nummedal's portrayal of Anna Zieglerin, secrets only became a problem when Zieglerin went overboard with the identity she created, while the seamstress Maddalena portrayed by Storey apparently sold her secrets for many years, despite several brushes with the law. Moreover, the specific cases of these two female purveyors are very different. Zieglerin enjoyed great favour at court for her alchemical secrets. Maddalena, in contrast, scraped by selling her magical and medical secrets for profit, aided by a male accomplice. While Zieglerin fell spectacularly and was subjected to a particularly gruesome execution, Maddalena likely escaped with at most a short prison sentence, as Storey notes that women accused of peddling magical cures were dealt with relatively leniently in Renaissance Venice.

In short, there were many different ways in which women could be involved with secrets, and while some played into longstanding stereotypes of women and secrecy, many others did not. In fact, as a number of the chapters in this volume demonstrate, it is very difficult to partition secrets definitively by gender. Mukherjee's study of Hugh Platt, for example, shows the extent to which Platt was interested in items usually considered part of housewifery: comfits and jams, cookery, household management. Similarly, the physicians' regimens described in Cavallo's chapter gave recommendations on such topics as household linens and pillows, perfumery, and combs for the hair and beard. These cases of male authors focusing on the secrets of domesticity call into question the perceived division in household roles and knowledge. More than simply expanding the demographic of actors interested in secrets to include women, then, the chapters in *Secrets and Knowledge* call into question perceived gender boundaries surrounding spheres of knowledge.

Conclusion

Secrets and Knowledge gathers together some of the most exciting current research on early modern secrets, but it is neither comprehensive nor definitive. The volume focuses most directly on medicine, alchemy and domestic science, the subjects represented in the printed books of secrets and analogous manuscripts. Yet secrets certainly played a key role in other sciences as well: astronomy, cartography, geography and navigation, to name a few. Another great lacuna is the

---

[43] Park, *Secrets of Women*; Susan Broomhall, *Women's Medical Work in Early Modern France* (Manchester: Manchester University Press, 2004) and Monica Green, *Making Women's Medicine Masculine: The Rise of Male Authority in Pre-Modern Gynecology* (Oxford: Oxford University Press, 2008).

world beyond Western Europe. Trade and colonial aspirations to the east and the west, a crucial source of commerce in secrets and the subject of much fascinating research, are sadly not represented here. Even within the parameters of this volume, there is great need for further research, as suggested in the opening chapters by William Eamon and Pamela H. Smith. While Eamon prompts researchers to conduct additional empirical studies with databases to track individual recipes and analyze their contents, Smith suggests that we work to construct and decipher the epistemologies behind these bits of know-how. Both these research schemas aim to achieve a deeper general understanding of the role secrets played in early modern knowledge transfer and production.

Secrets and recipes were ubiquitous all over Europe and beyond, but their role in knowledge making was shaped locally by a range of different factors. Literacy, politics, the structure and organization of households, the provision of materials objects within the home and the institutions regulating knowledge production all had an impact on their function. Needless to say, in early modern Europe, these factors differed greatly from country to country, town to town and village to village. The transnational nature of our volume, utilizing the expertise of a team of scholars, goes some way to explore these differences and similarities. Nevertheless, more research is urgently required – particularly in light of new studies on the commercial transactions between Europe and other parts of the world. We hope that this volume provides both a survey of the fascinating research currently being done in the field and a framework for carrying out new work on secrets and knowledge in early modern Europe.

PART I
Defining Secrets

Chapter 1
How to Read a Book of Secrets

William Eamon

As Venice's Carnival season of 1564 got underway, the poet Dionigi Atanagi sat down to write a dedication for publication in his friend Leonardo Fioravanti's latest book, *Capricci medicinali* (Medical Caprices). The book had so captivated him, Atanagi wrote, that he imagined himself staying at home reading his Fioravanti instead of going out to celebrate the Carnival gaiety:

> This Carnival, when others go to see masques, balls, and parties, as they are wont to do during this season, I'm staying at home with *The Medical Caprices* in my hands, reading it intently and to my great satisfaction: not only because it is a work by Your Excellency, whom I esteem with so much love, but much more, because of the excellent matters it treats and for the novel, quick, and safe way that you teach medicine and surgery.[1]

Fioravanti's new book, which was quite a sensation in its day, belonged to the genre of popular literature called 'books of secrets', vernacular works containing prescriptions, recipes and advice concerning medicine and other practical arts, including alchemy, cosmetics, metallurgy, perfumery, veterinary science and so on.[2] To anyone familiar with books of secrets, it must seem a little odd to imagine someone reading a recipe book with such sheer pleasure and enjoyment – indeed, to the point of foregoing the Carnival festivities. Of course, those familiar with the genre will also know that the books of secrets were not *merely* recipe books, but were also books that professed to reveal the arcane secrets of arts not even known to the authorities. Indeed, Atanagi went on at some length in this vein, calling Fioravanti's book 'a precious jewel' that embraced in just a few pages 'all that the immense volumes of Galen, Avicenna, and the other satraps of medicine could barely contain'. He lauded Fioravanti's plain style of writing, which, according to the poet, he accomplished 'with so much ease and simplicity of words, without dressing itself with smooth talk and artful rhetoric, attending only to the pure

---

[1] Letter dated 17 February, 1564, in Leonardo Fioravanti, *Capricci medicinali* (Venice, 1564), preface. All translations are mine.

[2] On Leonardo Fioravanti, see William Eamon, *The Professor of Secrets: Mystery, Medicine, and Alchemy in Renaissance Italy* (Washington, DC: National Geographic Books, 2010).

expression of the matter at hand, because he wants to be understood by the common people'.[3]

It may be difficult for modern scholars to fully appreciate Atanagi's way of reading a book of secrets, so accustomed are we to thinking of them as 'how to' books with purely didactic and utilitarian purposes. Yet his testimony bears witness to the fact that reading a recipe book for pleasure was one of several possibilities open to early modern readers. Contemporaries often read books of secrets quite differently from how authors imagined their audiences would read them, while historians read books of secrets for reasons that authors very likely never imagined anyone would. With these observations in mind, I would like to approach the question that I posed in my title – how to read a book of secrets – from three different but related perspectives:

The first is from the point of view of the authors who wrote books of secrets. I want to ask: How did they intend their books to be read and why did they choose to write in this particular genre? Since most early modern books of secrets were printed books, the question necessarily leads to a consideration of the profession of writing for the printing press and of printing for popular audiences.

The second perspective that I would like to consider is that of readers. How much can we know about who actually read books of secrets and how they read them? Beyond that, what light can these books shed on early modern popular and academic cultures?

The third perspective is that of modern historians. How have historians read books of secrets and how can reading books of secrets help us understand early modern culture? By way of orientation, I would like to begin by introducing two famous books of secrets whose simultaneous appearance in the year 1555 will provide the starting point for my discussion.

Secrets Occult and Revealed

In 1555, the last Renaissance edition of the Latin *Secretum secretorum* – the pseudo-Aristotelian 'Secret of secrets' that Aristotle supposedly communicated to his disciple, Alexander the Great – was published in Naples.[4] Its editor, Francesco Storella, a professor of rhetoric at the University of Naples, gave the work a good deal more attention and editorial care than had any scholar since Roger Bacon in the thirteenth century – certainly more, given the state of scholarship at the time, than it merited. For, by the time Storella's edition appeared, the *Secretum* was so generally regarded as spurious that sixteenth-century editors of the Aristotelian corpus rarely bothered to mention the work.

[3] Fioravanti, *Capricci medicinali*, preface.

[4] On the *Secretum*, see Steven Williams, *The Secret of Secrets: The Scholarly Career of a Pseudo-Aristotelian Text in the Latin Middle Ages* (Ann Arbor, MI: University of Michigan Press, 2003).

Although the *Secretum* continued to be popular in vernacular editions, to most humanists it was simply another example, among many so determined during the Renaissance, of a medieval text falsely attributed to a famous ancient author.[5]

That same year, the Venetian bookstalls displayed for the first time another book of secrets, one whose content and intended audience were entirely different from that of the *Secretum secretorum*. The book, *I Secreti del reverendo donno Alessio Piemontese* (The secrets of Alessio Piemontese), confidently advertised itself as 'a work very useful and universally necessary for everyone'.[6] Alessio's secrets included remedies unknown to the doctors, recipes for cosmetics used by the Turks, exotic perfumes and oils, dyeing techniques, tricks of the metalworking trades, and alchemical secrets tried out by Alessio himself. All of these precious 'secrets of nature', the product of his life's labour, Alessio stated, he was now publishing 'for the benefit of the whole world'.

Alessio's book of secrets was not a popularization in the usual sense; that is, it was not an adaptation of a text by one of the standard ancient or medieval authorities, such as Aristotle, Galen or Albertus Magnus. Rather, it was a completely new work – indeed, a new kind of work – and so was its author: in fact, its author was a fabrication. Although in his preface Alessio identified himself as a wealthy intellectual who had turned away from the books of the authorities in order to search for the 'secrets of nature', as far as we can tell, the *Secreti* was in fact the product of a Venetian popular writer by the name of Girolamo Ruscelli.[7] Writing under the pseudonym of Alessio Piemontese, Ruscelli created an enduring topos, that of the wandering searcher after the secrets of nature.

Alessio's *Secreti* was an instant bestseller. More than 100 early modern editions of the work were published, including translations into French, German, Dutch, English, Spanish, Polish and Latin.[8] In Italy the *Secreti* unleashed a torrent of books of secrets. So sensational were they in their day that in 1585, the Augustinian monk and social critic Tommaso Garzoni identified them as making up a new literary genre and their authors as a new profession. In his compendious *Piazza universale di tutte le professioni del mondo* (Universal plaza of all the

5 On this edition, see Charles B. Schmitt, 'Francesco Storella and the Last Printed Edition of the Secretum Secretorum', in W.F. Ryan and Charles B. Schmitt (eds), *Pseudo-Aristotle 'The Secretum Secretorum'* (London: Warburg Institute, 1982), pp. 124–31.

6 Alessio Piemontese, *I Secreti del reverendo donno Alessio Piemontese* (Venice, 1555). For discussion, see William Eamon, *Science and the Secrets of Nature: Books of Secrets in Medieval and Early Modern Culture* (Princeton, NJ: Princeton University Press, 1994), pp. 139–47.

7 On the authorship of the *Secreti*, see Eamon, *Science*, pp. 140–44.

8 For Spanish editions, see Mar Rey Bueno, 'Primeras ediciones en castellano de los libros secretos de Alejo Piamontes', *Pecia Complutense. Boletín de la Biblioteca Histórica de la Universidad Complutense de Madrid*, 2/2 (2005). Online: http://www.ucm.es/BUCM/foa/pecia/num02x/index02x.htm.

professions of the world), Garzoni called the new writers on secrets the 'professors of secrets' (*i professori de' secreti*), describing them as searchers after things 'whose reasons are not so clear that they might be known by everyone, but by their very nature manifested only to a few'. Although the professors of secrets sometimes make bombastic claims, Garzoni admitted, such secrets of nature were important because they 'contain certain seeds of discovery, which lead the intellect to finding out whatever it seeks to know'.[9]

No two works so similar in title could have been more different in content and purpose than these. Alessio's *Secreti* was unapologetically a book aimed at popular readers. It made no pretense of having academic merit. Storella's critical edition of the *Secretum secretorum*, on the other hand, was a work intended for scholars. Whereas the alleged author of the *Secretum* was the Prince of Philosophers himself, the sources of Alessio's secrets were not authors at all, but surgeons, empirics and miscellaneous 'experimenters' who had sought to improve the arts through trial and error methods. While Alessio's secrets consisted of practical recipes that anyone might use, pseudo-Aristotle's *Secretum secretorum* supposedly contained the wisdom that Aristotle reserved for his intimate disciples as opposed to the public doctrine that was studied in the schools. Although both were books of secrets, they advanced radically different ideologies concerning scientific discovery and the dissemination of knowledge. One decreed that the secrets of nature could be known only by revelation and were revealed only to those worthy of receiving them, while the other affirmed the value of long experience in the world and the utility of experiments that any skillful hand might manage. One defended an esotericist doctrine, while the other championed the ideals of empiricism and public science. As one tradition of secrets-literature came to a close in 1555, another opened.

Alessio's *Secreti* opened with a parable that became as much a part of the legend of Alessio as the recipes filling his book. In his preface, 'Don Alessio' reports that he was born into a famous noble family. Being independently wealthy, he was able to freely pursue his passion for 'philosophy and the secrets of nature'. Abandoning humanistic studies, he spent his entire life travelling from place to place collecting secrets. His sojourn took him all over Italy, to Germany, and twice to the Levant. He visited Jerusalem, Aleppo and Syria, and collected secrets, he says, 'not only from men of great knowledge and profound learning, and noblemen, but also from poor women, artisans, peasants, and all sorts of men'.[10]

Alessio's preface identified a secret's most valuable characteristic as a commodity, namely, its rarity. 'I've always been very miserly about wanting to communicate any of my secrets', he wrote. 'I've always said that if everyone

[9] Tommaso Garzoni, *La Piazza universale di tutte le professioni del mondo*, (ed.) P. Cherchi and B. Collina (Torino: Einaudi, 1996), p. 324.

[10] '...non solamente da grandi huomini per dottrina, & da gran Signori, ma ancora da pouere feminelle, d'artegiani, da contadini, & da ogni sorte di persone'. Piemontese, *Secreti*, fol. A2v.

knew the secrets, they wouldn't be secrets anymore, but just public and common things'.[11] Alessio's fame rested on his possession of rare secrets and he worried that publicizing them would cheapen them and thereby diminish his reputation. He paid for his vanity when, as he supposed, his refusal to give up one of his cherished secrets led to the death of a man who might have been saved had he communicated it to the attending surgeon. Gripped with remorse, he resolved to publish them all. Obviously, what Alessio and many others discovered was that, in the age of printing, publicity could be a more effective way to win fame than secrecy.

The Moral Burden of Secrecy in the Age of Printing

Now, the entire preface, as I have already indicated, was an invention and, like all stories, it carried a moral. What was Girolamo Ruscelli attempting to tell readers in fabricating this preface? How did he want readers to read his book? I would like to suggest that, in the first place, he was using the book of secrets as a vehicle to address a moral dilemma that faced all early modern writers who wrote for the popular press. Popular writers were continually challenged for 'prostituting' the secrets of the sciences by publishing translations of Latin works originally written for academic audiences and were challenged to defend their 'vulgarizations' against attacks by academics. Such polemics took place throughout Europe, wherever the popular press flourished. In mid-sixteenth century Germany, for example, a huge controversy erupted over the publication of popular medical tracts by writers such as Lorenz Fries and Walther Hermann Ryff. 'I was hated and persecuted by the learned physicians', Fries lamented, because I published the contents of this art in the German language'.[12]

Ruscelli himself became embroiled in a famous literary feud over popularization with the poet Lodovico Dolce when, in 1552, he and Dolce published rival editions of Boccaccio's *Decameron*. It was a sensational affair that had all of Venice talking. In a widely circulated pamphlet, Dolce ridiculed Ruscelli's 'paltry' efforts. Ruscelli shot back with a scathing review of Dolce's translation of Ovid, accusing Dolce of having demeaned the Roman poet's style. The duel intensified, and at one point in the quarrel, Dolce played on Ruscelli's reputation for dabbling in alchemy and magic, writing this diatribe:

> Everyone knows that Ruscelli is a good for nothing scoundrel, a swindler, and a cheat. He's ignorant and full of vice. Not being successful at alchemy, the pedantic

[11] '...sempre sono stato auarißimo di uoler communicare alcuno de'miei secreti ... Et sempre io diceua, che se i secreti si sapessero da ogn'uno, non si chiameriano più secreti, ma publici'. Ibid.

[12] Joachim Telle, 'Wissenschaft und Öffentlichkeit im Spiegel der Deutschen Arzneibuchliteratur', *Medizinhistorisches Journal*, 14 (1979): p. 34.

> profession that teaches all the learning of an ass, he has the temerity to translate Plutarch's *Lives* from Greek (of which he hasn't any more knowledge than a magpie), the Bible from Hebrew (which he knows about as well as my dog), and after a thousand ridiculous charlatan's recipes, he has lowered himself to the art of the pimp and has crammed his house with all sorts of courtesans and glad-handing prostitutes begging for the bread they can't earn by their own talents.[13]

If Alessio's preface was meant as a parable about the social danger of withholding specialized knowledge from the general public, it was also, by implication, a critique of the learned establishment. Alessio's English translator, William Warde, certainly took the work in that spirit, noting that, in the beginning, knowledge of the virtues that God had instilled in natural things was common to all men and only later concealed under the cloak of secrecy and learning. Warde, a Puritan, declared it his 'bounden duty' to reveal those secrets for the good of the commonwealth, not only to increase the public's health and welfare, but also to 'put the soul in remembrance of God's greater power and might, that hath given such virtue unto things growing on the earth, for man's commodity'.[14]

The futility and danger of esoteric knowledge – specifically the arcane alchemical disciplines – was also a prominent theme of Italian noblewoman Isabella Cortese's book, *I Secreti*. An avid alchemical practitioner, Cortese insisted that 'true alchemy' was technical alchemy, stripped of any mystical content. She was openly contemptuous of alchemical theory. Writing to her brother-in-law, a disappointed alchemical aficionado, she advised him not to bother reading the books of the famous alchemists:

> I tell you, dearest brother, if you want to follow the art of alchemy and to work at it, it isn't necessary to read the works of Geber, Ramon [Lull] or Arnaldo [Villanova] or any of the other philosophers, because they haven't recorded anything truthful in their books, just lies and riddles. … I have studied these books for more than thirty years and have never found anything good in them … Dearest brother, I know that you've wasted a lot of time and spent a lot of money. Therefore, out of compassion for you I beg you not to lose any more time on these books of the philosophers.[15]

---

13 Quoted in Paolo Trovato, *Con ogni diligenza corretto: La stampa e le revisioni editoriali dei testi letterari italiani (1470–1570)* (Bologna: Mulino, 1991), p. 241. The 'ridiculous charlatan's recipes' to which Dolce refers were those contained in Ruscelli's *Secreti nuovi* (see below).

14 Alessio Piemontese, *The Secretes of the Reuerende Maister Alexis of Piemount*, trans. William Warde (London, 1558), sig. *1v. Spelling modernized.

15 Isabella Cortese, *I Secreti … ne' quali si contengono cose minerali, medicinali, artificiose, & alchimiche, & molte de l'arte profumatoria, appartenenti a ogni gran signora* (Venice, 1574), pp. 19–21.

The denunciation of esoteric knowledge and the critique of the pretensions of traditional learning were themes that repeatedly surface in the early modern printed books of secrets. The professors of secrets saw traditional learning as a system of knowledge that was concealed under the cloak of the Latin language and obscure academic terminology. Bolognese surgeon Leonardo Fioravanti, one of the most prolific authors of books of secrets, praised the printing press for having overturned the reign of esotericism that gripped ancient medicine:

> In those days they got people to believe whatever they wanted, because there was a great shortage of books then, and whenever anyone could discourse even a little about *bus* or about *bas* he was revered as a prophet, and whatever he said was believed. But ever since the blessed printing press came into being, books have multiplied so that anyone can study, especially because the majority of them are published in our mother tongue. And thus the kittens have opened their eyes.[16]

To Fioravanti, the '*bus*' and '*bas*' of academic discourse kept the people in the dark and the physicians on top. The 'blessed printing press' changed all that; now, the people could learn formerly esoteric secrets on their own.

'Experimental' Authority and Public Personae

In 1567, after editing an additional three parts of Alessio's *Secrets*, Girolamo Ruscelli finally published a book of secrets under his own name. In the book, titled *Secreti nuovi* (New secrets), Ruscelli revealed that Alessio was a pseudonym that he had invented and that the secrets of Alessio were actually discovered experimentally in an academy that he had helped organize in Naples. Although Ruscelli's 'Academy of Secrets' (*Accademia Segreta*) met in secret and did not share their experiments with the wider world, they did so, Ruscelli reported, only in order not to attract attention to themselves: the group's intent, he insisted, was always, 'at the appropriate time' to reveal the secrets.[17]

On the face of it, Ruscelli's academy of secrets looks like an early scientific society and his book of secrets an 'experimental' book. However, I do not think that the authors of the books of secrets intended their works to be read as experimental books in the usual sense: that is, they did not expect, nor did they instruct, readers to experiment with the recipes or even to deviate from them in any way. The claim

16 Leonardo Fioravanti, *Dello Specchio di scientia universale* (Venice, 1567), fols 41r–v.

17 Girolamo Ruscelli, *Secreti nuovi di maravigliosi virtù* (Venice, 1567), fol. 5r. On Ruscelli's work and for an English translation of his preface, see William Eamon and Françoise Paheau, 'The Accademia Segreta of Girolamo Ruscelli. A Sixteenth-Century Italian Scientific Society', *Isis*, 75/2 (1984): pp. 327–42.

of the book of secrets is that the recipes had already been tried out – sometimes 'infinite times' – and therefore had already been certified by experiment.[18] In general, they did not encourage readers to experiment independently, only to follow the instructions exactly as written. Should you fail, it is because you did not follow the recipe to the letter.

Thus, instead of being prescriptions to experiment, as they are sometimes taken to be, the books of secrets provide rules to follow. In that sense, recipes are conservative: although they are prescriptions for taking action, for doing, they are premised on the assumption that the reader does not have the technical skills to make things up himself. 'Just follow the rules that I have written down for you', Isabella Cortese admonished her brother-in-law. 'Do not increase or diminish anything, but do what I say and write, and follow my commandments'.[19] Isabella emphatically did not encourage experimentation; she insisted that readers follow her recipes point by point, without deviating from the instructions in any way, just as you would when cooking from a cookbook.[20]

Hence, rather than serving primarily as guides to experimentation, I think that the books of secrets were meant to demonstrate the authority and expertise of the author. Eric Ash, who has written perceptively about the rise of a culture of expertise in early modern England, observes that the mark of the expert was 'his claim to mastery of some rare, valuable, and complicated body of useful knowledge'.[21] How do you demonstrate expertise? One way is to *lay bare* a process and spell out the secret.[22] By showing 'how to', you demonstrate that you command knowledge

---

[18] 'Experiment', as numerous historians have pointed out, is a problematic term in the early modern context. By and large, medieval and Renaissance philosophers used the terms 'experiment' and 'experience' interchangeably. However, the word *experimentum* ('experiment') gradually began to be used to refer to events lying outside the rational order, for example, in magic or manipulated events such as alchemical experiments. See Charles B. Schmitt, 'Experience and Experiment: A Comparison of Zabarella's View with Galileo's in *De motu*', *Studies in the Renaissance*, 16 (1969): pp. 80–138; Michael McVaugh, 'The *Experimenta* of Arnald of Villanova', *Journal of Medieval and Renaissance Studies*, 1/1 (1971): pp. 107–18.

[19] Cortese, *Secreti*, pp. 19–21.

[20] Of course, this does not mean that readers actually followed Cortese's admonition and never deviated from her instructions. Doubtless, readers ignored her commandment and altered the recipes in attempts to 'improve' upon them, even though Cortese disapproved of such practices, just as modern cooks often deviate from and experiment on cookbook recipes.

[21] Eric H. Ash, *Power, Knowledge, and Expertise in Elizabethan England* (Baltimore, MD: Johns Hopkins University Press, 2004), p. 9.

[22] For example, the case of Walther Hermann Ryff, in Alisha Rankin, 'How to Cure the Golden Vein: Medical Remedies as *Wissenschaft* in Early Modern Germany', in Pamela H. Smith, Amy Meyers and Harold Cook (eds), *Ways of Making and Knowing: The Material Culture of Empirical Knowledge* (Ann Arbor, MI: University of Michigan Press, forthcoming).

of the subject. The more complex or esoteric the process, the more convincing is the expertise, because being an expert in something difficult is more impressive than being an expert in something simple and easy to accomplish.

For whom were such demonstrations of expertise and authority intended? First of all, for patrons and potential patrons: that is, they aimed to show potential patrons that one had something useful to offer. Leonardo Fioravanti, one of the most famous professors of secrets identified by Garzoni, used his works as a way of fishing for patrons, dangling his books in front of various nobles and princes in the hope of converting them to his 'new way of healing'. The 'new way' was as much as anything a clever advertising gimmick, which Fioravanti touted as being radically different from the 'old way'. A doctrine premised on the idea that all diseases stem from a single cause, bodily corruption engendered in the stomach, the theory had radical implications. Indeed, it amounted to the claim that the doctrine of the four humours was the Great Lie of the physicians. For if all illnesses stemmed from a single cause, then all could be cured by a single universal medicine, regardless of the patient's individual humoural complexion. It was a theory that perfectly matched the marketplace's demand for fast and cheap cures. Fioravanti's 'new way of healing' and the brand names that he gave his panaceas captured the mood of the times. He dedicated works to Cosimo d'Medici, Alfonso II, the Duke of Ferrara, King Philip II of Spain, and others, often using the same dedication for several prospective patrons. He never succeeded in finding a patron to support him, although he was briefly employed in the court of Philip II. He had better luck on the popular front.[23]

In fact, there is reason to believe that Fioravanti's main aim was not to garner patronage but to craft a public image. He was a genius at it. He used the prefaces and dedications to his books – multiple dedications for every book – as a way of associating himself with princes, patricians and leading cultural figures and to hone his public image as a potent healer. Spectacular episodes of conflict – such as the alleged 'conspiracy of the doctors' that drove him from Rome – lent an aura of sensationalism to his life story, which he published in his popular *Tesoro della vita humana* (Treasury of human life).[24] Chris Rojek, in his book, *Celebrity*, refers to this kind of performative tactic as 'staged celebrity', in other words, 'calculated technologies and strategies of performance and self-projection designed to achieve a status of monumentality in public culture'.[25] Fioravanti, like few others in his day, grasped the importance of the media in self-presentation and in the formation of public opinion. By exploiting the possibilities of early modern printing, he fashioned a persona and self-image as a celebrity healer.

23 Eamon, *Professor of Secrets*, esp. pp. 207–9.

24 Leonardo Fioravanti, *Il Tesoro della vita humana* (*Treasury of Human Life*) (Venice, 1570), Book III.

25 Chris Rojek, *Celebrity* (London: Reaktion Books, 2001), p. 121. Rojek uses the example of James Boswell, the biographer of Johnson, who projected an image of himself as a celebrity biographer.

Marketing Secrets

There is no question that Fioravanti was a brilliant popular writer. His works – eight in all – went through a combined total of 90 early modern editions and translations, making him almost as popular an author as Alessio Piemontese. In fact, just as he had discovered a 'new way of healing' that differed from the ordinary physicians' methods, he professed to have invented an entire 'new way of writing, different from that of all other authors'. He wrote:

> Just as in the fish market they sell all kinds of fish – including expensive fish for rich people and less costly kinds for the poor – so that everyone can eat according to their rank and not go hungry, so it's necessary to have all sorts of writers. Some write for high and exalted minds, some for the middling sort, and some for those who don't understand very much at all. In this way there will be food for all. Therefore if my work is not for intellectuals and men of learning, nor even for those of middling quality, at least it will be for those who understand little. For they are the hungriest and I want everyone to have some food for thought.[26]

Frankly, this is hard to believe. Fioravanti's recipes hardly seem intended for people 'who don't understand very much at all'. In fact, they are extremely complex and often require ingredients and equipment that only a specialist, such as a pharmacist or an alchemical adept, would have access to. The measurements are often exacting, requiring a scale or specialized measuring utensils. Most of his drugs were made by distillation and he was quite specific in requiring that the distilling apparatus be made of glass and not metal, which, he claimed, can corrupt the results. These were not everyday household recipes; they were complex proprietary compounds that carried catchy trade names such as Leonardo's Grand Liquor, Angelic Electuary, Magistral Syrup, and *Dia Aromatica*, the 'Fragrant Goddess' that he prescribed as the first course of action against almost every ailment he encountered. And, he assured his readers, if you didn't want to bother to make the drugs yourself, you could buy them ready-made at the Bear Pharmacy in Venice.

If Fioravanti's drugs were made for sale at the pharmacies, why did he go to the trouble of publishing the recipes? He did so, I suggest, in order to place a proprietary mark on them. At a time when there was no effective pharmaceutical patent system, inventors of drugs that they hoped to market in multiple cities – as Fioravanti did – used books of secrets to assert their property rights. The strategy seems to have worked. One of the Venetian owners of his popular *Capricci medicinali* was the Dominican friar and distiller Antonio Volpe, who was tried for heresy in 1567. We know about the friar's library – just four books, including a breviary, a psalter and a copy of Fioravanti's *Medical Caprices* – from his trial before the Venetian Inquisition. Questioned about the books, Volpe testified that

---

[26] Fioravanti, *Capricci medicinali*, p. 1.

none were heretical and that the *Capricci* was a reference work that he used in his practice.[27] Fioravanti's patent drugs, which were available not only in Venice but also in pharmacies in Naples, Rome, Milan and London became a standard part of the pharmacist's stock. His drugs were still found in the Italian pharmacies as late as the 1780s and were widely used both within and outside of Italy. The message that Fioravanti projected in his books – touting remedies that were invented by a man of experience and tested by experiment – would not have been lost on general readers. Even if he lacked the skills required to duplicate the recipes, the reader of a book of secrets could be confident that the author did have them. In the hands of a clever entrepreneur like Fioravanti, the book of secrets served as an effective means to advertize and market drugs.

Fioravanti was by no means unique in using the book of secrets as a vehicle for marketing drugs. In his recent study of charlatanism in early modern Italy, David Gentilcore notes that charlatans not only marketed their drugs on the piazzas but also through the recipe books that they sold, thus expanding their business beyond the piazza and to more settled locations. For example, when the Brescia surgeon and distiller Andrea Fontana was licensed to sell his secret for the stone in Siena, he made specific mention of his 'printed recipe collection'. Competition was fierce on the piazzas, and the recipe book may have been used as a way of gaining a competitive advantage over other charlatans. By its inclusion in a printed recipe book, Fontana's secret gained a kind of certification of its efficacy.[28]

Thus, the emphasis on experiment and practice in the book of secrets was not necessarily intended to encourage experimentation on the part of casual readers. Rather, the book of secrets was a mechanism for asserting the validity of natural knowledge gained by practice. This was something quite new. Pamela H. Smith, in her recent book, *The Body of the Artisan*, has argued persuasively that a new kind of epistemology – what she calls an 'artisanal epistemology' – emerged in the early modern period, claiming that knowledge of nature can be gained by a bodily encounter with natural things.[29] Beyond empiricism, beyond mere observation, the artisanal epistemology was premised upon knowing by manipulation and engagement with matter. This kind of knowledge comes not from reading books but from doing.

The books of secrets occupy an interesting place within this scheme of knowledge. The professors of secrets, I think, subscribed to an epistemology very

---

[27] On Volpe, see William Eamon, 'The Canker Friar: Piety and Intrigue in an Era of New Diseases', in F. Mormando and T.W. Worcester (eds), *Piety and Plague in Europe: From Antiquity to the Early Modern Period*, (Kirksville, MO: Truman Statue University Press, 2007) pp. 156–76.

[28] David Gentilcore, *Medical Charlatanism in Early Modern Italy* (Cambridge: Cambridge University Press, 2007), p. 363. I discuss some of these pamphlets in *Science and the Secrets of Nature*, pp. 239–44.

[29] Pamela H. Smith, *The Body of the Artisan: Art and Experience in the Scientific Revolution* (Chicago, IL: University of Chicago Press, 2004), p. 59.

much like the artisanal epistemology that Smith described. However, emboldened by the opportunities for self-fashioning opened up by the printing press, they took that claim about knowledge one step further, asserting that natural knowledge gained by doing can, in fact, be communicated through printed recipes. Although a recipe cannot substitute for knowledge gained by practice, it is a guide for replicating craft-like processes, providing a route to knowledge that cannot be gained by traditional means. William Warde, the English translator of Alessio's *Secreti*, justified his work on the grounds that everyone 'by a certain instinct or natural inclination, desires to know things not known before and to hear news not before heard'. He invited readers not only to 'take pleasure in the knowledge' of Alessio's secrets, but, 'if it pleases you, to put into experience and prove those that seem best to you'.[30] By following the recipes and replicating the processes, he implies, one satisfies one's desire for natural knowledge.

Reading and Using Books of Secrets

I do not mean to diminish the importance of the didactic intent of the authors of books of secrets. As William Warde's preface implies, books of secrets were also intended to be used as instructional manuals, and many were doubtless read that way. Certainly they were meant to be useful. Did readers find them so? How did they use them? Even if Fioravanti's books were too complicated to serve as household recipe books, that was by no means true of all books of secrets. Many of the English copies of 'Alexis of Piedmont's' *Secrets* that I have examined contain marginal comments indicating that the book was not only read but used. Even if some writers of books of secrets – Isabella Cortese and Leonardo Fioravanti, for example – discouraged readers from deviating from their instructions, readers did not shy away from experimenting with ingredients and procedures, substituting ingredients, changing the amounts specified, and even pronouncing them useless if their experiments found them so. One particularly sceptical reader of a 1580 edition of the *Secrets* deleted numerous passages from the text, noting that 'All theas receipts ar verye falsly written, but being corrected heer they ar trew'.[31] Many copies of the text exhibit extreme wear and, doubtless, many copies were worn out of existence. Some contain underlining, corrections and additional recipes, sometimes personal and sometimes copied out of other texts, handwritten in margins and back pages. A much-used specimen of the 1562 edition of the *Secrets* preserved in the Regenstein Library at the University of Chicago contains extensive marginal commentary, including additional recipes, corrections to the text and readers' judgments such as 'very good'. Alessio's recipes are quite

[30] Piemontese, *Secretes*, sig. *1r.

[31] William H. Sherman, 'What Did Renaissance Readers Write in Their Books?', in Jennifer Andersen and Elizabeth Sauer (eds), *Books and Readers in Early Modern England: Material Studies* (Philadelphia, PA: University of Pennsylvania Press, 2002), p. 126.

accessible and susceptible to replicating in the kitchen, so to speak. It is clear from the marginal comments preserved in many copies of the text that readers did not simply replicate recipes, but changed, tested and 'corrected' them.[32]

Similarly, Isabella Cortese's book, although pitched at alchemical adepts, was actually quite popular, going through seven sixteenth-century Italian editions, and contains recipes that could have been managed in a kitchen laboratory. Its eclectic blend of recipes – from metallurgical and alchemical processes to perfume and hair dyes, and from spot-removing to counterfeiting pearls that look just like natural ones – gave the book a comprehensive, encyclopedic heft; and the claim that the recipes were tested by an experienced alchemist might have enhanced its appeal among general readers.

Even more accessible as household reference books were the little pamphlets that went under the title, *Opera nova intitolata Dificio di ricette* (Palace of Recipes), which were published continuously in Italy from the 1520s and spun off numerous imitations.[33] This little booklet can perhaps be best described as a general domestic 'how to' book, since, as the title page notes, it contains 'many very useful recipes, which will also be pleasing'. Besides the usual mix of recipes typically found in books of secrets – making soap, perfumes, cosmetics and so on – *Dificio di ricette* includes various medieval methods to determine whether a wife or husband is impotent, along with parlour tricks and magical illusions, such as making a candle burn under water and a ring to dance on the table. To early modern readers, there was something wonderful about such diversions, and this may offer a clue to understanding Atanagi's fascination with Fioravanti's *Medical Caprices*.

Dificio di ricette and countless books like it raise the question: What distinguishes a 'book of secrets' from a mere recipe book? On the face of it, very little. In terms of content, there is in fact little that distinguishes this body of texts from the books of secrets, except that they were smaller and perhaps less pretentious. On closer inspection, however, there are some important differences. For one thing, alchemy (in particular, distillation) typically occupies a more prominent place in the books of secrets than in household recipe books. This is not to imply that alchemy is absent from the household recipe books.[34] Yet, for the professors of secrets, alchemy was not just a tool but a means to *experiment on nature*; which leads to a second difference between the two genres: Books of secrets typically represented themselves as serving natural philosophical ends as well as utility. In other words, 'secrets' tended to be regarded as clues leading the investigator to uncover the *secrets of nature*, or, in Garzoni's words, as 'seeds of

---

32 William H. Sherman's studies of marginalia in early modern texts are especially important in helping us understand how readers read their books. See William H. Sherman, *Used Books: Marking Readers in Renaissance England* (Philadelphia, PA: University of Pennsylvania Press, 2008).

33 *Opera nova intitolata Dificio di ricette* (Venice, 1529). For a preliminary inventory of this body of writings, see Eamon, *Science*, pp. 361–5.

34 Indeed, the *Dificio di ricette* contains a woodcut drawing of a friar working at a still.

knowledge' that can lead to new discoveries.[35] However, one should perhaps not place too much stress upon such distinctions. There can be little doubt that readers used books of secrets in multiple ways.

Beyond serving as practical recipe books, the books of secrets were useful in another important way: namely, they gave readers the assurance that nature was predictable and that craft and professional secrets were not beyond their ken. As Alison Kavey writes, books of secrets presented readers with models of a natural world that is 'susceptible to human manipulation'.[36] This kind of knowledge was useful to early modern readers for a number of reasons. First of all, knowing that the mysterious forces of nature are predictable and to some extent rational and orderly was empowering to readers because it meant that the secrets of nature were within one's intellectual grasp. Scholars have stressed that forces of secularization were at work in the early modern era to transform European society from one dominated by clerical and aristocratic elites to one in which laymen and burghers held a significant share of wealth and political power.[37] The sphere of religion was diminished, so that many of the hopes and fears formerly expressed in religious terms came more and more to be expressed in worldly terms. William Bouwsma has written, 'Secularization rested on a deep conviction that eternal truths are inaccessible to the human intellect, and that only the limited insights afforded by experience in this world are relevant to the earthly career of the human race'.[38] The books of secrets, which compressed the experience of generations of empirics into simple, time-tested rules, were among the most visible manifestations of such attitudes. The explosion of printed 'how to' books publicizing the secrets of the trades and the tricks of magic made it clear as never before that a recipe might effectively replace the artisan's cunning, and that what had formerly passed as magic could now be seen as mere hocus-pocus and sleight of hand.[39]

In addition, familiarity with the secrets of the crafts could make you a more intelligent consumer, an important asset in an emerging commercial economy. Urban dwellers in early modern Europe were surrounded by strange and exotic commodities, often from places they'd never heard of. The marketplace, increasingly populated by itinerant merchants and healers who were here today and gone tomorrow, symbolized concerns about the city as a moral and economic environment. The issue of fraud and knowing whom to trust in the marketplace

---

35 William Eamon, 'Science as a Hunt', *Physis*, 31/2 (1994): pp. 393–432.

36 Alison Kavey, *Books of Secrets: Natural Philosophy in England, 1550–1600* (Urbana, IL: University of Illinois Press, 2007), p. 3.

37 Natalie Zemon Davis, *Society and Culture in Early Modern France* (Stanford, CA: Stanford University Press, 1975).

38 William J. Bouwsma, 'The Secularization of Society in the Seventeenth Century', in *A Usable Past: Essays in European Cultural History* (Berkeley and Los Angeles, CA: University of California Press, 1990), p. 114. In addition, see Eamon, *Science*, p. 130.

39 See, for example, John White, *Hocus-Pocus* (London, n.d.), where the 'hocus-pocus' of magic is spelled out in simple recipes.

became critically important.[40] These everyday concerns are mirrored in popular writings about mountebanks and in the widespread concerns about alchemical frauds that Tara Nummedal has written about.[41] What is camphor root from Borneo used for? How do you judge its quality and how do you prepare it? How do you choose the best lignum aloes? Is balsam of Peru a good substitute for ancient balsam? Laying bare the mysteries of the trades could enable readers to answer these questions, and to better understand how craftsmen made the things they bought. Even if one did not necessarily have any interest in replicating the recipes they encountered in a 'how to' book, they still might find the secrets useful in daily life.

Books of secrets could also teach consumers how to successfully negotiate their way in a changing medical marketplace. In a recent essay, Elaine Leong and Sara Pennell observe, 'Being a "smart consumer" in the mixed medical economy of the period also meant being prepared to engage in an informed fashion in that economy, and protect oneself from exploitation'. Books of secrets and manuscript recipe collections could help consumers avoid 'the predations of an overenthusiastic apothecary' or, for that matter, a charlatan.[42] The diversity of treatment options available to consumers in early modern Europe – from orthodox physicians and apothecaries to wise women and charlatans – gave the sick more choices, but also complicated the choices and made the issue of trustworthiness more critical than ever.

I agree, then, with Alison Kavey that books of secrets constructed a certain kind of reader, a reader with a somewhat changed position in relation to the natural world. I once (perhaps a little too enthusiastically) labelled the sixteenth century the 'age of how to' on the grounds of the growing tide of books of secrets.[43] Even if that was a stretch, I think that there is something to be said for the idea that books of secrets contributed to removing the veil of mystery from nature. Some readers of the *Dificio di ricette* might have been disappointed by the results they experienced from trying out the prognostication instructions and magical tricks they encountered in the book. On the other hand, lack of success may have encouraged more sceptical attitudes towards magic, especially when compared to the results of the more strictly technological recipes they found in the work. Readers may not have understood *why* certain materials hardened iron or made

40 Katharine Park, 'Country Medicine in the City Marketplace: Snakehandlers as Itinerant Healers', *Renaissance Studies*, 15/2 (2001): pp. 104–20.

41 Tara Nummedal, 'The Problem of Fraud in Early Modern Alchemy', in Richard Raiswell and Mark Crane (eds), *Shell Games: Scams, Frauds and Deceits in Europe, 1300–1650* (Toronto: University of Toronto Press, 2004), pp. 37–51.

42 Elaine Leong and Sara Pennell, 'Recipe Collections and the Currency of Medical Knowledge in the Early Modern "Medical Marketplace"', in Mark S.R. Jenner and Patrick Wells (eds), *Medicine and the Market in England and its Colonies, c. 1450–c. 1850*, (London: Palgrave Macmillan, 2007), p. 146.

43 Eamon, *Science*, p. 126.

fast dyes, but they had a better idea of *how to do it.* The books of secrets gave answers to the timeless question, 'How in the world did he do that?'

It is difficult to say how seriously contemporary readers took the lofty moral counsel that they would have encountered in books of secrets like Alessio's *Secreti.* Did the narrative that was constructed by the professors of secrets about the moral imperative of 'going public' and freely revealing secrets to all have a wide appeal? It must have. Otherwise, it is hard to see why the narrative was so widely imitated. If Alessio Piemontese was a worldly seeker who travelled widely in search of experiences, learning from all sorts of people, so too were Leonardo Fioravanti, Isabella Cortese and John Hester, the London apothecary and Fioravanti's English translator.[44] The world-weary professor of secrets who, as Fioravanti wrote of himself, 'walked the world and ploughed the seas' in search of the secrets of nature, and who freely revealed his hard-won secrets to the public, was a topos that had widespread appeal.[45] Even the charlatans posed as intrepid seekers of secrets from distant parts who would freely disclose their secrets in the little chapbooks. One charlatan, who called himself 'Benedetto the Persian', sold a pamphlet titled *I Maravigliosi et occulti secreti naturali* (The Marvellous and Occult Secrets of Nature), a booklet of secrets supposedly 'translated from Persian into Italian'.[46] Licensed for sale in multiple cities, the pamphlet deliberately appealed to the Renaissance fascination with the exotic. If printers knew their audiences – and surely they did – they must have known that the moral tone of the book of secrets resonated with many readers.

Books of Secrets and the History of Science

I began my discussion by asking three questions about how to read a book of secrets. The last concerns historiography: How should historians read books of secrets? What is the historical significance of the wave of books of secrets that streamed from the presses in early modern Europe?

The first modern historian to write seriously about books of secrets was the University of Glasgow chemist and bibliographer John Ferguson. Ferguson (1837–1916), who was Regius Professor of Chemistry at Glasgow, was an avid book collector whose personal library numbered more than 7,500 books on alchemy, chemistry and the occult sciences. In 1896, he began presenting a series of papers on books of secrets to the Glasgow Bibliographical Society. His addresses, collected in *Bibliographical Notes on Histories of Inventions and Books of Secrets* (1896–1910), constitute the first bibliography of the genre. Ferguson's

---

[44] On Hester, see Deborah Harkness, *The Jewel House. Elizabethan London and the Scientific Revolution* (New Haven, CT and London: Yale University Press, 2008), pp. 93–6.

[45] Leonardo Fioravanti, *Tesoro della vita humana*, fol. 1r.

[46] Benedetto 'Il Persiano', *I Maravigliosi et occulti secreti naturali*, (Rome, Venice, Bologna, Milan, 1613).

Bibliographical Notes is not often cited or, evidently, read anymore.[47] That is a pity because, besides identifying books of secrets as a genre, he said some shrewd and learned things about them. Even so, in the light of recent historiography, we have to admit that he comes across as a little old-fashioned. To Ferguson, books of secrets were relics of an outmoded world of chemistry just one remove from (as he saw it) the dark, arcane world of the medieval alchemist. In his very first essay on books of secrets, first published in 1896, he wrote, 'With the progress of the sciences and physical change, books of secrets … can no longer show any reason for existence. If they circulate at all it is as chap books, the hawkers' reprints of Aristotle, Albertus Magnus, etc., which have little interest from a bibliographical, and none from a scientific point of view, or else as collections of trivial receipts which are of no practical use'.[48]

Of course, to modern historians, such a description is an invitation to look closer. We as historians do not read books of secrets that way anymore, namely, as evidence of how infinitely better off we are now that we rely on sound and rational scientific theory as opposed to simple empiricism and rule of thumb. However, I mention Ferguson because it is really not so long ago that historians did read them that way – if they read them at all. When I gave my first professional paper on books of secrets to the History of Science Society in 1982, in a panel on the Scientific Revolution, the presentation was met with blank stares of puzzlement and deadly silence. No one since Ferguson (whose final essay on the subject was published in 1910) had written about books of secrets, and hardly anyone in the audience could understand what they could possibly have to do with the history of science. Perhaps it was the quality of the paper, certainly a work in progress; but it was a depressing moment. So disappointing was the experience that I actually set the project aside for several years before returning to it.

Thanks to the turn towards cultural history of science that has characterized the discipline since then, and to an entirely new way of understanding the role of alchemy, the craft tradition and manual work in the Scientific Revolution, no one needs to seek any kind of validation to study books of secrets anymore. To the current generation of historians, the idea of 'making knowledge' and the investigation of how scientific knowledge is constructed are key areas of concern.[49]

---

47 John Ferguson, *Bibliographical Notes on Histories of Inventions and Books of Secrets*, collected edition (Staten Island, NY: Pober Publishing, 1998). On Ferguson and his important collection, see David Weston, 'A Magus of the North? Professor John Ferguson and his Library', in Amy Wygant (ed.) *The Meanings of Magic: From the Bible to Buffalo Bill* (Oxford: Berghahn Books, 2006), pp. 161–77.

48 Ferguson, *Bibliographical Notes*, I-20. The article was originally a lecture given to the Archeological Society of Glasgow in 1882.

49 See, for example, Smith, *Body of the Artisan* and the essays in Lissa Roberts, Simon Schaffer and Peter Dear (eds), *The Mindful Hand: Inquiry and Invention from the late Renaissance to Early Industrialisation* (Amsterdam: Koninklijke Nederlandse Akademie van Wetenschappen, 2007); and in Pamela H. Smith and Benjamin Schmidt (eds), *Making*

In those days, however, the professional discipline of the history of science was still tied to the classic works in the history of science, to intellectual history and to the problem of the 'rise of modern science', although, thanks to voices like those of Charles Webster, Michael Hunter and a few others, even then the field was being slowly pulled away from those seemingly secure moorings.

As a result of these changes, the landscape of the Scientific Revolution looks completely different from the way it looked a generation or two ago. No longer is alchemy considered to be one of history's dead ends or as an obstacle to development experimental science, but as an important part of its history.[50] Recent research has also given us an entirely new understanding of the importance of the artisanal tradition for the development of science. 'Making knowledge', that is, the process by which knowledge is generated, has become as important as the end product of knowledge making – books, ideas and so on – in the narrative of the origins of modern science.[51] In this new scheme of things, as this volume makes clear, books of secrets have become an important part of the narrative of the Scientific Revolution. Moreover, recent studies of books of secrets have played an important role in reshaping that narrative. All of this bodes well for the discipline, although there is much more work to be done.

What Can Historians Learn from Books of Secrets?

In conclusion, I would like to pose a few questions and suggest some problems and hypotheses for further exploration of this fascinating subject. Recently, historians have called attention to a growing interest in medical and natural philosophical particulars in the Renaissance.[52] Descriptive knowledge based upon the cataloguing of particular facts became increasingly important to natural philosophers, in contrast to knowledge based upon universals, which was characteristic of medieval natural philosophy. A kind of knowledge that was in the Middle Ages considered to be confused, random and unstable became, in the early modern period, the foundation of a new natural philosophy. Historians including Lorraine Daston, Katharine Park, Nancy Siraisi and Alisha Rankin have provided

Knowledge in Early Modern Europe: Practices, Objects, and Texts, 1400–1800 (Chicago, IL: University of Chicago Press, 2007).

[50] For a perceptive discussion of the historiography of alchemy in the Scientific Revolution, see William R. Newman, *Atoms and Alchemy. Chymistry and the Experimental Origins of the Scientific Revolution* (Chicago, IL: University of Chicago Press, 2006).

[51] For a survey of such new approaches, see Pamela H. Smith and Benjamin Schmidt, 'Knowledge and its Making in Early Modern Europe', in Smith and Schmidt, *Making Knowledge*, pp. 1–16.

[52] For an overview and extensive bibliography, see Nancy Siraisi and Gianna Pomata (eds), *Historia: Empiricism and Erudition in Early Modern Europe* (Cambridge: Cambridge University Press, 2005).

excellent signposts for this line of inquiry.[53] Although the explosion of books of secrets has been acknowledged as a sign – among numerous other signs – of this growing interest in particulars, it seems to me that pursuing this line of inquiry further might prove fruitful. There is much to be learned, I think, by a closer examination of the recipes that fill the books of secrets.

This will not be a simple task. One of the frustrations confronting the scholar who studies books of secrets is the very profusion of recipes that characterizes the genre. It can seem daunting. Yet it seems to me that one of the things that is so greatly needed is exactly what many of us have avoided doing: actually pouring through books of secrets and parsing each recipe by ingredients, processes, use, mode of employment and so on, rather than simply thinking and writing about books of secrets as a genre. In other words, we need to look closer at the particulars. Databases of recipes are needed. Lines of transmission need to be drawn. Connecting such research to the growing early modern interest in natural particulars might yield surprising results.

The social anthropologist Jack Goody, one of the most perceptive commentators on the recipe as a medium for communicating knowledge, suggested that there is a continuum from the recipe to the prescription to the experiment.[54] Recipes are, in effect, prescriptions for taking action, which not only extends the repertoire of specialists but also, inevitably, leads to comparative testing: in other words, to experimenting. Whether intended that way or not, books of secrets probably did lead some readers to experiment. Perhaps that is what Garzoni meant when he said that secrets contain 'seeds of knowledge' that lead to new discoveries. Thus, it seems to me, Goody's hypothesis still has relevance. I still do not think that we fully understand the significance of books of secrets for the rise of experimental science.

How exactly are recipes related to experiments and what light can books of secrets shed on the emergence of experimental methodologies? Peter Dear observed that when the Fellows of the Royal Society of London reported their experimental findings, they frequently employed a recipe-like format in their communications.[55] The reader was given a specific set of instructions for an experimental procedure, which he could replicate on his own. It would be interesting to know more about the

[53] Examples include: Lorraine Daston, 'The Factual Sensibility', *Isis*, 79/3 (1988): pp. 452–67; Katharine Park, 'Natural Particulars: Medical Epistemology, Practice, and the Literature of Healing Springs', in Anthony Grafton and Nancy Siraisi (eds), *Natural Particulars: Nature and the Disciplines in Renaissance Europe* (Cambridge, MA: MIT Press, 1999), pp. 347–67; Alisha Rankin, 'Becoming an Expert Practitioner: Court Experimentalism and the Medical Skills of Anna of Saxony (1532–85)', *Isis*, 98/1 (2007): pp. 23–53.

[54] Jack Goody, *The Domestication of the Savage Mind* (Cambridge: Cambridge University Press, 1977), Ch. 7.

[55] Peter Dear, '*Totius in verba*: Rhetoric and Authority in the Early Royal Society', *Isis*, 76/2 (1985): pp. 145–61.

origins of this methodology. Books of secrets would have provided the community of experimenters with a ready-made formula for such expressions. Indeed, the English virtuoso Robert Boyle was an avid reader of books of secrets and mined them for information. Yet he remained sceptical about their contents unless the recipes were actually tested experimentally. Boyle warned readers not to trust alchemical experiments that are communicated only by way of prescriptions and not of 'relations': 'that is, unless he, that delivers them, mentions his doing it upon his own experience'.[56] Fine-grained studies of books of secrets – perhaps even experimental studies like the ones that William Newman is doing for Newton's alchemy – might help us uncover new sources of early modern experiments.[57]

Recipes are among the most ancient of literary forms. One of the earliest medical texts in the western tradition, the Ebers Papyrus, consists of a collection of recipes for treating various eye, skin and internal ailments.[58] Thousands of recipes exist in medieval and early modern manuscripts.[59] They have been studied from the angles of the history of medicine, the history of technology, the history of the arts and crafts, and the history of science. Yet we know surprisingly little about how they were actually used. It would be helpful to have more studies along the lines of Jennifer Stine's unpublished Stanford dissertation on books of household medicine in England, and of Elaine Leong's equally impressive studies of both manuscript and printed household medical collections.[60] The process by which household medicine, long buried in manuscript, was discovered by printers and popular writers is a fascinating story. Of course, simply writing down a recipe or printing a book of recipes does not mean that the recipe was actually used. But occasionally the manuscripts reveal surprising testimony about the uses of household medicine and technology.

The study of these manuscripts provides an entry into the history of everyday life and the history of therapeutics, still the least understood part of the history

---

[56] Robert Boyle, *The Sceptical Chymist*, in Thomas Birch (ed.) *The Works of the Honourable Robert Boyle*, (6 vols, Hildesheim: Georg Olms Verlag, 1966), vol. 1, p. 460.

[57] William Newman, *The Chymistry of Isaac Newton* (2005–2009), Indiana University. Online: http://webapp1.dlib.indiana.edu/newton/ [accessed: 2 April, 2009].

[58] Reinhold Scholl, *Der Papyrus Ebers. Die größte Buchrolle zur Heilkunde Altägyptens*, Schriften aus der Universitätsbibliothek, 7 (Leipzig, 2002).

[59] Gerhard Eis, *Mittelalterliche Fachliteratur*, 2nd edn (Stuttgart: J.B. Metzler, 1967).

[60] Jennifer K. Stine, *Opening Closets: The Discovery of Household Medicine in Early Modern England* (PhD diss., Stanford University, 1994); Elaine Leong, *Medical Recipe Collections in Seventeenth-Century England: Knowledge, Gender and Text* (D.Phil. thesis, University of Oxford, 2005); Elaine Leong, 'Making Medicines in the Early Modern Household', *Bulletin of the History of Medicine*, 82 (2008): pp. 145–68. In addition, see Sara Pennell, 'Perfecting Practice? Women, Manuscript Recipes and Knowledge in Early Modern England', in Victoria Burke and Jonathan Gibson (eds), *Early Modern Women's Manuscript Writing: Selected Papers from the Trinity/Trent Colloquium* (Aldershot: Ashgate Publishing, 2004), pp. 237–58.

of medicine. This way of reading books of secrets, it seems to me, holds much promise. What are we to make, for example, of the magical tricks and amusements that populate booklets like the *Dificio di ricette*? By spelling out the secrets of magic and showing how to do the trick, did such booklets have the effect of making the world of magic seem a little less magical? Finally, the fact that many of the household recipe books were composed by women opens up a fascinating chapter, which scholars have only recently begun to explore, in the history of gender.[61] Research has revealed, among other things, a substantial and growing number of literate women in sixteenth and seventeenth century Europe, and that women relied on books of secrets and other 'how to' books to learn technologies (for example, needlework) formerly taught solely by oral tradition.[62] Read carefully, the household recipe books can also yield much information about the sphere of women's work, the organization and supervision of domestic spaces, and the wide range of technical knowledge skill required to manage early modern households.[63] Moreover, as Lynette Hunter and others have shown, there is much to be learned from household recipe books about the involvement of women in experimental activities.[64]

Books of secrets turn up in a surprisingly wide variety of libraries, from those of ordinary middle class readers to specialists in the crafts to alchemists to princes. Duke Cosimo d'Medici was an avid collector of books of secrets, while King Philip II of Spain had access to many titles in his great library at El Escorial. For modernizing princes concerned with the welfare of their realms, books of secrets joined all sorts of technological manuals and treatises in providing useful information. We even encounter books of secrets in the libraries of minor noblemen like the Spanish patrician Vincencio Juan de Lastanosa, who lived in the remote town of Huesca, in Aragon. For collectors like Lastanosa, books of secrets made up an important part of the cabinet of curiosities.[65]

---

61 See, for example, Alisha Rankin, 'Duchess, Heal Thyself: Elisabeth of Rochlitz and the Patient's Perspective in Early Modern Germany', *Bulletin of the History of Medicine*, 82 (2008): pp. 109–44; Emanuela Renzetti, 'La Sessualità nei libri dei segreti del XVI e XVII secolo', in Gisela Bock and Giuliana Nobili (eds), *Il Corpo delle donne* (Bologna: Transeuropa, 1988), pp. 47–68.

62 Elizabeth Tebeaux, 'Women and Technical Writing, 1475–1700: Technology, Literacy, and the Development of a Genre', in Lynette Hunter and Sarah Hutton (eds), *Women, Science and Medicine 1500–1700: Mothers and Sisters of the Royal Society* (Stroud: Sutton Publishing, 1997), pp. 29–62.

63 Lynette Hunter, 'Women and Domestic Medicine: Lady Experimenters, 1570–1620', in Hunter and Hutton, *Women, Science and Medicine*, pp. 89–107.

64 Lynette Hunter, 'Sisters of the Royal Society: The Circle of Katherine Jones, Lady Ranelagh', in Hunter and Hutton, *Women, Science and Medicine*, pp. 178–97. In addition, see Rankin, 'Becoming an Expert Practitioner'.

65 William Eamon, 'Appearance, Artifice, and Reality: Collecting Secrets in Courtly Culture', in Mar Rey Bueno and Miguel López-Pérez (eds), *The Gentleman, the Virtuoso,*

The apparently wide distribution of the genre leads to interesting questions having to do with the history of popular culture and the origins of secularism. Did technical literacy contribute in some measure to the 'disenchantment of the world' that Max Weber noted was one of hallmarks of modern culture?[66] It is difficult to know. Perhaps the avalanche of self-help manuals showing readers how to draw and paint pictures, assay metals, dye fabrics, make gunpowder, and handle all manner of tools and instruments contributed to a sense of independence on the part of many who became self-taught. Did books of secrets play a role in the rise of the culture of expertise? Did they educate consumers and make them more intelligent buyers, as I suggested earlier? Were they used as sources by the virtuosi as they sifted valid empirical data from spurious testimony? We still do not have conclusive answers to these questions, and they are questions that are still worth asking.

Many of the recipes in the books of secrets fell under the category of alchemy – even making glue and soap were considered alchemical. Only occasionally did the professors of secrets engage in the great debate of the day over the question of true versus false alchemy, a subject that Tara Nummedal has explored with fascinating results in the German context.[67] That debate took place all over Europe, and it seems clear that most professors of secrets would have sympathized with the view expressed by Isabella Cortese, who did engage in the debate. Eschewing gold-making and false alchemical promises, the professors of secrets advanced a more modest, result-oriented kind of alchemy that became a model for the 'true alchemy' espoused by Boyle, Andreas Libavius, and others.[68] If these hypotheses have merit, we might find a larger role for books of secrets in the narrative of the Scientific Revolution and the origins of modernity.

Finally, returning to the point that I made at the outset, when I contrasted two famous books of secrets, I would like to suggest that books of secrets have something important to tell us about changes in the moral economy of early modern science.[69] Alessio Piemontese's conversion to the ethic of openness, his

the Inquirer: Vincencio Juan de Lastanosa and the Art of Collecting in Early Modern Spain (Cambridge: Cambridge Scholars Publishing, 2008), pp. 127–43.

66 Max Weber, 'Science as a Vocation', in H.H. Gerth and C. Wright Mills (trans. and ed.), *From Max Weber: Essays in Sociology* (New York, NY: Oxford University Press, 1946), pp. 129–56; Weber, *The Protestant Ethic and the Spirit of Capitalism*, trans. Talcott Parsons (1930; repr., London: Routledge, 1992), p. 61.

67 Tara Nummedal, *Alchemy and Authority in the Holy Roman Empire* (Chicago, IL: University of Chicago Press, 2007).

68 Lawrence M. Principe, *The Aspiring Adept: Robert Boyle and His Alchemical Quest* (Princeton, NJ: University of Princeton Press, 2000); Bruce T. Moran, *Andreas Libavius and the Transformation of Alchemy* (New York, NY: Science History Publications, 2007).

69 I make this argument in 'The Secrets of Nature and the Moral Economy of Early Modern Science', in A. Paravacini Bagliani (ed.), *Il Segreto/The Secret*, *Micrologus*, vol. XIV (Florence: SISMEL, 2006), pp. 215–35.

rejection of medieval esotericism, and his dramatic announcement of his intention to reveal his precious 'secrets of nature' to the general public, whatever the cost of doing so, symbolizes a critical realignment of values within the moral economy of science. Renouncing esotericism and proclaiming his conversion to the ethic of public knowledge, Alessio dedicated himself to pursuing hidden secrets of nature, verifying them by experiment and publicizing them to the world at large. His preface, widely reprinted in many different languages and editions, helped to create the topos of the early modern professor of secrets. Numerous Renaissance writers, from Paracelsus to Leonardo Fioravanti, exploited the topos as a way of self-fashioning in an age when *personae* were created in printed books. In retrospect, Alessio's most important contribution to early modern science was not his compilation of experiments but his discovery that publishing secrets was ethically superior to concealing them from the unworthy. In the age of printing, that reversal would be decisive.

Nevertheless, in spite of the vocal denunciation of esotericism that issued from the pens of the professors of secrets, there continued to be much resistance to open disclosure of craft, medicinal and experimental 'secrets' down through the seventeenth century. In 1665, Robert Boyle addressed the issue in his very first publication, a pamphlet titled *An Epistolical Discourse Inviting all True Lovers of Vertue and Mankind, to a Free and Generous Communication of their Secrets and Receits in Physick.*[70] In the *Discourse*, Boyle articulated a powerful argument against secrecy and esotericism in science. Not only did Christian charity demand open disclosure, he argued, but the sciences would advance much further if secrets were made public and subject to experimental testing. Yet with respect to revealing secrets for the general good, even Boyle was, in the words of Michael Hunter, a 'reluctant philanthropist'. Indeed, he was 'riven with anxiety about the advisability of making available material which he was convinced it was both his Christian duty, and in the public interest, to disseminate'.[71] He worried that disseminating recipes might damage his reputation if any were found to be false, and employed a standard literary technique of alchemical authors, dispersion, whereby he willfully separated laboratory processes into fragments and sprinkled them randomly throughout his writings.[72] Although Boyle has often been portrayed as a champion of open communication of scientific knowledge and a consistent opponent of all

70 The tract was printed in a collection of works edited by Samuel Hartlib, *Chymical, Medicinal, and Chyrurgical Addresses* (London, 1655). Boyle's tract was published by Margaret E. Rowbottom, 'The Earliest Published Writing of Robert Boyle', *Annals of Science*, 6/4 (1948/50): pp. 376–87.

71 Michael Hunter, 'The Reluctant Philanthropist: Robert Boyle and the "Communication of Secrets and Receits in Physick"', in Ole Peter Grell and Andrew Cunningham (eds), *Religio Medici: Medicine and Religion in Seventeenth-Century England* (Aldershot: Ashgate Publishing, 1996), p. 265.

72 Lawrence M. Principe, 'Robert Boyle's Alchemical Secrecy: Codes, Ciphers and Concealments', *Ambix*, 39/2 (1992): pp. 63–74.

forms of secrecy, his much-touted 'free communication' of recipes turned out in practice to be much more problematic than he imagined when he wrote his tract condemning secrecy.

* * *

Everyone loves a secret and everyone has a secret, which is why books of secrets continue to fascinate us. On the face of it, books of secrets are the most transparent of documents. What could be simpler than a recipe? What could be more transparent than a 'how to' book? Yet, as any scholar who has seriously studied them knows, early modern books of secrets are by no means transparent. In fact, they present the serious reader with many puzzles about why they were written, for whom, and how they were read. In addition, there is such a great variety of books of secrets, from household recipe books to professional manuals to works that served as vehicles for advertising, that they are not easily categorized. I am reminded of Robert Darnton's comment about encountering a joke that you don't get: when historians realize they are not getting something – a joke, a proverb, a ceremony – Darnton observed, they have located a starting point into an otherwise opaque aspect of a culture.[73] As long as we as historians, in reading books of secrets, continue to squint in bemusement or puzzlement, we can rest assured that we are on to something.

[73] Robert Darnton, 'Workers Revolt: The Great Cat Massacre of the Rue Saint-Séverin', in *The Great Cat Massacre and Other Episodes in French Cultural History* (New York, NY: Basic Books, 1984), pp. 77–8.

Chapter 2
What is a Secret? Secrets and Craft Knowledge in Early Modern Europe

Pamela H. Smith

Varieties of Secrets

'Secrets' conjures up mysterious, perhaps marvellous knowledge, and many books of 'secrets' gesture towards marvels and wonders or allude to the esoteric nature of their contents. The fourteenth-century *Book of Secrets* ascribed to Albertus Magnus, for example, is accompanied by a 'book of the marvels of the world', while the thirteenth-century *Book of the Secrets of Alchemy* by Constantine of Pisa begins with an explanation of the title: 'secrets which are greatly confused and known to very few. But let no one wonder at that, for knowledge must be concealed, lest it be demeaned: pearls must not be cast before swine; and he who divulges mystic secrets diminishes their divinity'.[1] Secrets were also advertised by charlatans and medical practitioners when they set up shop, displayed the 'tricks of their trade' and sold their remedies on public squares.[2] Secrets thus connoted wondrousness and concealment, perhaps even deception, but despite these connections with spectacle and the spectacular, most books of secrets are filled with more or less straightforward recipes. As Constantine's *Book of the Secrets of Alchemy* continues: 'These secrets can be entitled: congealing, inhuming, boiling, cleansing, washing, subliming, distilling, annealing and thus making perdurable',

---

1 Albertus Magnus (attributed), *The Book of Secrets of Albertus Magnus of the Virtues of Herbs, Stones and Certain Beasts: Also A Book of the Marvels of the World*, ed. Michael R. Best and Frank H. Brightman (Oxford: Clarendon Press, 1973); Constantine of Pisa, *The Book of the Secrets of Alchemy*, trans. and comm. B. Obrist (Leiden and New York, NY: Brill, 1990), p. 227.

2 The Jesuit Giovan Domenico Ottonelli recorded the show put on by charlatans: 'On occasion a company of these gentlemen [charlatans] appears in town … They spread the word that they want to serve the public, selling excellent secrets and putting on good entertainments …'. He goes on to describe the selling of the secret by a 'main actor', then after 'having done a good trade in it and collected the money, the main selling comes to an end'. Then a comedy is performed. Giovan Domenico Ottonelli, Christiana Moderatio del Theatro, libro ditto l'ammoniamento a' Recitanti (Florence, 1652), p. iv, 455, cited in David Gentilcore, '"Charlatans, Mountebanks and Other Similar People": The Regulation and Role of Itinerant Practitioners in Early Modern Europe', *Social History*, 20/3 (1995): pp. 297–314 (p. 309).

in other words a set of operations known to any metalworker or distiller.[3] Even when the recipes employed impossible ingredients, such as basilisks, or impossible techniques, such as plunging a rock crystal into the still-beating heart of a he-goat, the recipes have a straightforward, often naturalistic tone and sound like the fruit of experience. Indeed many recipes note that they are 'tried' or 'experienced': for example, Constantine's *Book of the Secrets of Alchemy* begins, 'In this truthful and authoritative book, tried and tested through numerous experiments [carried out] by Master Constantine in accordance with the sayings of the philosophers, especially Aristotle's'.[4]

These books of secrets fall within the category of 'maker's knowledge', as William Eamon noted some time ago. Although they may make reference to an esoteric ideal of knowledge, they emerged in early modern Europe to have more in common with a modern day technical instruction book or, better still, a cookbook.[5] Indeed, it is telling that by the sixteenth century 'secrets' and 'arts' seem to be synonymous, as for example, in the 1616 German translation of the pseudonymous Alessio Piemontese's book of secrets that was titled *Book of the Art* (Kunstbuch) *of the Experienced Mr. Alexis Piedmontese, about Many Useful and Valuable Secrets or Arts*.[6] In the 1604 'Goldsmith's Storehouse', an assayer's book of techniques, an anonymous goldsmith author sees 'sciences' and 'mysteries' as synonymous, stating that a 'grounded experience in this Science or mysterie' is necessary to the practitioner of the craft.[7] By the eighteenth century, a collection of recipes gathered from various books of secrets could be titled a 'laboratory', as in Godfrey Smith's *The Laboratory: or, School of Arts* (1770).[8] A guild might enjoin its members to hold close the

3 Constantine of Pisa, *Book of the Secrets*, p. 227.

4 Ibid.

5 William Eamon, *Science and the Secrets of Nature: Books of Secrets in Medieval and Early Modern Culture* (Princeton, NJ: Princeton University Press, 1994), describes well the genesis of books of secrets out of esoteric ideals of knowledge and craft recipe books. A somewhat different category of 'secrets' that will not be treated here is discussed extensively in Katharine Park, *Secrets of Women: Gender, Generation, and the Origins of Human Dissection* (New York, NY: Zone Books, 2006). There were of course many sorts of secrets other than technical ones. See the essay review of Eamon's *Science and the Secrets of Nature* by Silvia De Renzi, 'Secrecy, Power and Knowledge in Early Modern Italy', *Studies in the History and Philosophy of Science*, 27/3 (1996): pp. 397–407.

6 *Kunstbuch des Wohlerfarnen Herrn Alexii Pedemontani, von mancherleyen nutzlichen und bewerten Secreten oder Künsten jetzt newlich auß Welscher und Lateinischer sprach ins Teutsch gebracht*, trans. Hanβ Jacob Wecker, Stadartz zu Colmar (Basel: Ludwig König, 1616).

7 The knowledge and craft ordinances of English gold- and silversmiths were known as the 'mystery' of the craft. 'The Goldsmith's Storehouse' [c. 1604], Folger Shakespeare Library, Washington, DC, MS V.a.179, fol. 6v. Another copy of this ms. is owned by Goldsmith's Hall.

8 Godfrey Smith, *The Laboratory: Or, School of Arts* (5th edn., London, 1770).

secrets of the trade – as was proclaimed by a general meeting of masons' lodges in the German-speaking lands in 1459: 'Also no workman, nor master, nor parlier, nor journeyman shall teach anyone, whatever he may be called, not being one of our handicraft and never having done mason work, how to take the elevation from the ground plan'[9] – or even include ritualized 'mystery' in its initiation procedures.

However, such secrecy did not necessarily refer to a hidden body of knowledge (in the case of the German masons, one of their number published this secret within a generation), but instead could be formulaic and might even be employed as a cover to hide officially prohibited processes such as subcontracting.[10] As Anne-Françoise Garçon makes clear, such practice of 'secrecy' was part of the culture and socialization of craft as well as a means of holding on to the position of expert, and came to be opposed to the openness of knowledge proclaimed by the writers of the *Encyclopedie* and other Enlightenment authors.[11]

What is a Secret? The Articulation of Craft Experience

Books of secrets, then, were an articulation of the experiential knowledge of craftspeople and practitioners that was 'hidden' in the things of nature or in the material objects they made, or hidden from the understanding of those not

---

9 Paul Frankl, 'The Secret of Medieval Masons', *The Art Bulletin*, 27/1 (1945): pp. 46–60 (p. 46). In 1486, Matthäus Roriczer published the secret in a short booklet (*Buch der Filialen Gerechtigkeit*). Roriczer was probably present at the meeting, although he did not sign the stipulations (p. 47). Although he published the 'secret', no one except a mason could have seen that his text presented a special case of the generalized method of producing a plan of elevation from a ground plan.

10 Anne-Françoise Garçon, 'Les Dessous des Métiers: Secrets, Rites et Soustraitance dans la France du XVIIIe Siècle', *Early Science and Medicine*, 10/3 (2005): pp. 378–91. See also the excellent discussion of the partly ad hoc and invented nature of the 'mysteries' of the French *compagnonnages*, more or less informal associations of French journeymen which appear to have arisen in the seventeenth century in response to a weakening of the legal divisions among the particular trades in which a guild member could engage. Michael Sonenscher, 'Mythical Work: Workshop Production and the *Campagnonnages* of Eighteenth-Century France', in Patrick Joyce (ed.), *The Historical Meanings of Work* (Cambridge: Cambridge University Press, 1987), pp. 31–63.

11 Earlier writers such as Cipriano Piccolpasso also desired to extract craft knowledge from craftsmen in the name of the public good. In *The Three Books of the Potter's Art* (ca. 1558), Piccolpasso declares that he is making public the art of ceramics against the 'ill-will of those in whose hands these secrets have been' so that it would move from 'persons of small account, [to] circulate in courts, among lofty spirits and speculative minds'. Cipriano Piccolpasso, *The Three Books of the Potter's Art* (ca. 1558), trans. R. Lightbown and A. Caiger-Smith (2 vols, London: Scholar Press, 1980), vol. 2, pp. 6–7.

experienced in the craft.[12] This articulation in an ideal of esotericism, in so far as the learning of a craft involved initiation, often initiation both into a social group and a technical body of knowledge that was not kept 'secret' or deliberately concealed so much as was not necessarily expressible in words and not codified in writing. In early modern Europe, craft 'secrets' were to a large degree tacit and ineffable. They involved techniques and practices that were often too complicated to be written down but instead could only be learned by observing, attending, imitating and doing. As such, they might be compared to the ineffable knowledge of the religious mystic. The mystery of union with God could not be expressed in rational form, but instead might be experienced through the body. The Muslim mystic Abu Hamid Muhammad ibn Muhammad al-Ghazali (1058–1111) stated that true knowledge cannot be expressed in words but can only be experienced as a kind of 'tasting'. Similarly, medieval Christian mystics, especially female mystics, described union with God through the language of bodily sexual union, hence the extraordinarily erotic – to our ears – language of some medieval religious writings. Sensory apprehension – tasting, smelling, seeing, feeling and even hearing God – gave a kind of certain knowledge that words and propositional knowledge could not. This was a knowledge known in the body, and, like craft knowledge, was often referred to as 'secrets' or 'mysteries'. The artisan had to sound out his materials, to be attuned to them, to taste them through the bodily senses, or to 'overhear' matter, as the medical and intellectual reformer Paracelsus in the sixteenth century expressed it in trying to capture this element of artisanal practice.[13]

While this view emphasizes the ineffable and mystical quality of craft knowledge, secrets can also represent a purely practical advantage in the marketplace, as John K. Ferguson in *Bibliographical Notes on Histories of Inventions and Books of Secrets* indicated:

> It is hardly to be expected that a practical art can have any literature worth speaking of. The man who is busy practicing it can have little time to write

[12] Vannoccio Biringuccio, *The Pirotechnia*, trans. C. Stanley Smith and M. Teach Gnudi (New York, NY: Basic Books, 1943), speaks about the hidden virtues of natural things that can be discovered through 'art': Biringuccio believed that all things created by God have some particular gift or power, and if humans cannot discern it, it is because of our 'defective vision, in our little knowledge, and in our lack of careful thought concerning the necessity of seeking hidden things. Certainly those things that have such inner powers, like herbs, fruits, roots, animals, precious stones, metals, or other stones, can be understood only through oft-repeated experience. Others do not need so much knowledge and belief for their powers are obvious and shown to us at all times by their effects. They reveal themselves to us through the sense of sight by their splendor and variety of colors, through the sense of smell by their odors, through the sense of hearing by the melody of the harmonious or unpleasantly discordant repercussions of the air. They likewise manifest themselves by their softness or harshness', p. 114.

[13] See Walter Pagel, *Paracelsus: An Introduction to Philosophical Medicine in the Era of the Renaissance*, 2nd edn (Basel: Karger, 1982), p. 51.

> about it, and he who wishes to learn it must put to his hand and work at it, and that under the supervision of a master, and not by merely reading books. This is the apprenticeship that everyone must serve. No amount of reading will make a sculptor, or a gardener, or a shoemaker, or a surgeon, or a musical executant. The arts must be acquired by practice, and they are extended and improved by practice. Everyone who exercises them comes to have special power and certain ways of doing things, which may enable him to surpass others who are similarly engaged. These are his 'secrets', which very often he cannot or will not, reveal to others. Rapid insight into a particular case, power of overcoming physical obstacles, ingenious adaptations of means to ends, exhibition of due care at the right time, enable one man to effect what others cannot.[14]

Until recently, craft secrecy has often been viewed as about the protection of technical secrets. However, in the last decade economic historians have shown that, while such 'secrecy' might give the practitioner an economic advantage for a short time, there is little evidence of processes actually being kept secret for very long and much evidence to the contrary showing just how rapidly technical innovations spread.[15] Indeed in an eighteenth-century Habsburg investigation of whether to retain the journeyman *Wanderschaft* among craft guilds, the report from the Duchy of Krain emphasized that journeymen could learn much that was useful to the territory and would introduce such innovations when they returned: 'it is known that in this way very frequently many secrets are discovered and new arts and techniques are introduced'.[16] Even the closely held 'arcanum' of porcelain making did not take long to spread from Meissen throughout Europe.[17] As Larry Stewart has noted, 'The power of guilds could not lie in jealously guarded secrets. It lay in the collective willingness of tradesmen to maintain standards and control over commerce'.[18]

---

14 John K. Ferguson, *Bibliographical Notes on Histories of Inventions and Books of Secrets* (2 vols, 1898, repr. London: The Holland Press, [1959]).

15 See Reinhold Reith, 'The Circulation of Skilled Labor in Late Medieval and Early Modern Central Europe', in Stephan R. Epstein and Maarten Prak, *Guilds, Innovation, and the European Economy, 1400–1800* (Cambridge: Cambridge University Press), pp. 114–42, esp. p. 137; and Karel Davids, *The Rise and Decline of Dutch Technological Leadership: Technology, Economy, and Culture in the Netherlands, 1350–1800* (2 vols, Leiden: Brill, 2008), vol. 1, p. 235 discussing the ineffectiveness even of Venice's bans on the transfer of knowledge from the Murano glassworks.

16 'nachdeme es bekannt, daß auf diese Arth sehr offt viele Geheimnisse entdecket, neue Künsten und Arbeiten eingeleitet', quoted in Reinhold Reith, 'Know-How, Technologietransfer und die *Arcana Artis* im Mitteleuropa der Frühen Neuzeit', *Early Science and Medicine*, 10/3 (2005): pp. 349–77 (p. 364).

17 Ibid., pp. 372–5.

18 Larry Stewart, 'Science, Instruments, and Guilds in Early Modern Britain', *Early Science and Medicine*, 10/3 (2005): pp. 392–410 (p. 408).

Writing Secrets Down

The articulation of craft experience had been codified in writing since ancient times and, while scattered writings, often having been translated and compiled from the Arabic, appeared in Europe during the Middle Ages, there was an explosion of vernacular European technical writing beginning around 1400, which was extended still further by the invention of printing in the 1460s.[19] Recipe collections and technical treatises on metalworking, dyeing and distillation were some of the earliest works off the presses and some of the biggest bestsellers, growing larger with each reprinting, as material was added and pseudonymous names changed. Early works in the vernacular appear in astounding numbers, including distillation techniques (*Distillir-Bücher*) in the 1490s, *Kunstbüchlein* (*Little Books of Art*) in the 1500s, assaying and metalworking treatises (known as *Probir-Büchlein*) in the 1530s. Alessio Piemontese's *Book of Secrets* burst onto the scene in the mid-sixteenth century, first in Italian then quickly in many other vernacular European languages as well as in Latin, going through 90 editions by the seventeenth century. Piemontese's book of secrets contained recipes for medicines, conserves, cosmetics and distilled 'waters', dyes and pigments, artificial precious stones and metalworking. This boom in 'how-to' books continued throughout the seventeenth century, which saw the publication of books on every conceivable subject, from beehive construction to composting and from embroidery to cannon casting.[20] The books were frequently compiled by printers and often varied significantly from edition to edition.

Such books of recipes and secrets have not been studied as a genre, but rather have generally been treated separately in terms of their subject matter: cookbooks, medical recipes, gunners' manuals, nautical guides, metalworking texts, books of 'secrets' and so on.[21] But if we study the wide range of practical guides – or better

---

[19] On one of the most influential of these Arabic books of secrets, translated in the twelfth century into Latin as the *Liber vaccae* or *Book of the Cow* from the now almost entirely lost ninth-century Arabic version, see Maaike van der Lugt, '"Abominable mixtures": The *Liber Vaccae* in the Medieval West, or the Dangers and Attractions of Natural Magic', *Traditio*, 64 (2009): pp. 229–77. Recipes from this collection continued to be included in books of secrets and alchemical treatises into the sixteenth century, including that attributed to Albertus Magnus cited above. Its contents also occasioned much scholastic debate about the legitimacy of natural magic and even informed controversies about witchcraft.

[20] For a survey of some of this literature, see Natasha Glaisyer and Sara Pennell (eds), *Didactic Literature in England 1500–1800: Expertise Constructed* (Aldershot: Ashgate, 2003). Alison Kavey also treats this literature, in *Books of Secrets: Natural Philosophy in England, 1550–1600* (Urbana, IL: University of Illinois Press, 2007). For an excellent introduction, see Jo Wheeler, *Renaissance Secrets, Recipes and Formulas* (London: V&A Publishing, 2009).

[21] A very important exception to this generalization is the work of Pamela O. Long, whose *Openness, Secrecy, Authorship: Technical Arts and the Culture of Knowledge from Antiquity to the Renaissance* (Baltimore, MD: Johns Hopkins University Press, 2001),

said, 'technical writing' – which appear from the late fourteenth century on in Europe as a connected phenomenon, several questions arise, above all: who were they for? What relationship did they have to everyday practice of the crafts? And what were they meant to convey or teach? A typical 'secret' or recipe described a set of ingredients or operations in more or less detail, but we should not assume from this that recipes were solely meant to provide 'how to' information. Although they might be employed for the practical information they conveyed, Elaine Leong, Sara Pennell and Alisha Rankin have made clear that they also possessed social functions and exchange value. As Leong and Pennell note about medical recipe collections,

> Therapeutic and pharmacological information and the ways in which it was disseminated, judged and validated can be seen as sharing some of the characteristics of currency, as it was understood by contemporaries: a commodity which flowed between people, and the authority and reliability of which was inflected by the circumstances of that movement ... recipes can be seen as analogous to particular forms of early modern financial transaction, notably bills of exchange, in that their realizable value was tied up with the trustworthiness of the relationship on which the exchange was based. But recipe exchanges also at times involved recipes as a variety of gift, where the values placed on the texts donated and received were framed by social relations, as much as any inherent 'value' in the recipe itself.[22]

Indeed, collections of recipes and techniques run the gamut from straightforward instructions, through advertisement for a practitioner's abilities, to wonder-working promise. Allison Kavey in *Books of Secrets: Natural Philosophy in England, 1550–1600* argues that, in most cases, these books offered readers not actual experience but rather the sense that nature was accessible to and manipulable by everyman (and, in the case of some subjects, by everywoman).[23] The appearance of these technical guides has also been associated with the growth of urban culture and the

---

was groundbreaking on several fronts, but most notably in giving insight into the multiple textual traditions of technology from antiquity through the sixteenth century.

22 Elaine Leong and Sara Pennell, 'Recipe Collections and the Currency of Medical Knowledge in the Early Modern "Medical Marketplace"', in Mark Jenner and Patrick Wallis (eds), *Medicine and the Market in England and its Colonies c. 1450–c. 1850* (London: Palgrave Macmillan, 2007), pp. 133–52 (pp. 133–4). See also Alisha Rankin, 'Becoming an Expert Practitioner: Court Experimentalism and the Medical Skills of Anna of Saxony (1532–1585)', *Isis*, 98/2 (2007): pp. 23–53.

23 Allison Kavey, *Books of Secrets.* On women's technical writing, see Elizabeth Tebaux, 'Women and Technical Writing, 1475–1700: Technology, Literacy and Development of a Genre' in Lynette Hunter and Sarah Hutton (eds), *Women, Science and Medicine 1500–1700: Mothers and Sisters of the Royal Society* (Thrupp: Sutton Publishing, 1997), pp. 29–62, although her assumption that technical writing arose as a result of the demise of the oral tradition cannot be correct.

increase in urban centres of a 'middling sort', who in their social mobility were more isolated from familial sources of technical knowledge and more desirous of new information that might be useful in their emulation of their social betters.[24]

While much more research needs to be undertaken on the vexed question of who made up the audience for technical writing, in what follows, I examine a number of recipes and recipe books to begin to investigate the relationship between craft practice and practical guides, and what these books were intended to convey. I will argue that in the apparently random jumble of recipe collections, a vernacular 'philosophy' of nature and a set of guideposts for thinking about and approaching matter can be discerned. One dimension of this technical writing was thus an attempt to put into writing the tacit, experiential knowledge of craftspeople. This knowledge included both theoretical frameworks and directives about attitudes and techniques. We might think about these *Kunstbücher*, these books of art, as trying to convey the essence of *Kunst*. This term should not be mistaken for 'art' in the sense of the fine arts, but must be sought for in its origins in *können* – 'to be able to do' – for these books were about the ability to 'do', to make and to produce. A *Kunststuck* meant a masterpiece, a trick, a secret, something that was not obvious to an individual who did not know the tricks of the trade. These books of art dealt with the quintessentially human capability to produce by handwork.

Varieties of Recipes

Let us begin with a collection of recipes for pigment-makers and dyers written down in the fifteenth century in Old Middle German known as *The Strassburg Manuscript*. A typical recipe is for a bright blue dye:

> to make a lovely bright blue rag colour for dying yarn, for a transparent glaze over silver, for letters in blue – 'About a week after Whitsun, get up early in the morning, at sunrise so as to be able to get in time to where the flowers are growing. Then pick handfuls of these blue flowers – and this must be before midday because after that they are no good. You will need three or four persons to break the tops off into a clean pot and as soon as that is done…'[25]

[24] A thorough study of these practical guides would have to include conduct manuals, exemplified by Baldassare Castiglione's *The Courtier* (1528) and imitated in numerous vernacular texts, as well as the *Hausväterliteratur* which seems to have been aimed at managers of estates. Rudolf Bell, *How to Do It: Guides to Good Living for Renaissance Italians* (Chicago, IL: Chicago University Press, 1999), examines guides for married couples published in the sixteenth century. In other words, a complete study of this genre would need to be quite expansive.

[25] *The Strassburg Manuscript: A Medieval Painters' Handbook*, trans. V. and R. Borradaile (London: Alec Tiranti, 1966), p. 37.

Many of the recipes in this collection are straightforward and can apparently be taken as instructions.

When we turn to other pigment-making and metalworking techniques and recipes, many of the instructions are not as straightforward. In the medieval goldsmith and monk Theophilus's thirteenth-century collection of metalworking techniques, most instructions could easily be followed by a goldsmith today, but a few are not so practical.[26] For example:

> If you want to carve a piece of rock crystal, take a two- or three-year-old goat and bind its feet together and cut a hole between its breast and stomach, in the place where the heart is, and put the crystal in there, so that it lies in its blood until it is hot. At once take it out and engrave whatever you want on it, while this heat lasts. When it begins to cool and become hard, put it back in the goat's blood, take it out again when it is hot, and engrave it. Keep on doing so until you finish the carving. Finally, heat it again, take it out and rub it with a woollen cloth so that you may render it brilliant with the same blood.[27]

As noted above, there is a striking specificity about these unlikely techniques – why a two-to-three-year-old goat or a woollen cloth, for instance? The recipe strives for a tone that indicates that the practice comes out of experience even when it in fact goes back millennia more or less unchanged to at least Pliny.[28]

Recipes for softening hard substances and hardening soft substances or the related process of fixing volatile materials are to be found in overwhelmingly large numbers in such collections. Some do cause changes of state, such as dissolution in acid, or heating, or even tempering of steel, but many of them, such as this goat's-blood recipe, are repeated over and over again, despite the fact they could not possibly have functioned as intended. It may be that the origin of this recipe was to crack crystals in order to allow dye penetration.[29] However, it seems more likely that such recipes instead are reiterated because they reinforce the philosophical framework of the Hippocratic–Galenic–Aristotelian view of nature in which opposites must be combined (or 'tempered') in order to bring the four elements/humours/qualities into a balance that results in good health. It is important to note that materials we now think of as inanimate were not regarded as such in the pre-industrial world. Minerals and metals grew in the earth and they had to be purged and tempered in order to bring them into a healthy state. Books of secrets

---

26 As Erhard Brepohl has shown in his *Theophilus Presbyter und die Mittelalterliche Goldschmiedekunst* (Vienna: Bohlau, 1987).

27 Theophilus, *The Various Arts. De Diversis Artibus*, ed. and trans. C.R. Dodwell (Oxford: Clarendon Press, 1986), pp. 189–90.

28 Pliny of course consulted works available to him, probably including the original of the Greek *Papyrus Graecus Holmiensis*, also known as the *Stockholm Papyrus*, which contains a recipe for softening crystals with goat's blood.

29 Kurt Nassau, *Gemstone Enhancement* (London: Butterworths, 1984), p. 10.

often indiscriminately mix medical recipes for human health with instructions for pigment, dye and metallurgical recipes. This should not be seen as a random combination, for all these recipes operate on the basis of the same coordinates of the four humours and qualities and the same principles of tempering in order to bring about balance (and thus health). This is supported by Albertus Magnus's comments about goat's blood in which he notes that the diamond – by which he means three different types of stones (diamond, magnetite, very hard metal) – cannot be softened with fire or iron, but it can be destroyed by the blood and flesh of a goat, 'especially if the goat has for a considerable time beforehand drunk wine with wild parsley or eaten mountain fenugreek; for the blood of such a goat is strong enough even to break up a stone in the bladder, in those afflicted with the gravel'.[30] Albertus moves easily from the stones to be worked by an artisan to those to be cured by a physician.

Although these recipes might refer back to a textual tradition, they were not static. The recipes were not copied mindlessly, but rather, authors and collectors made choices among them. Although some recipes might have originated in the distant past (which may have made them appear more trustworthy to their copyists), they were often combined with apparently firsthand experience.[31] Indeed the goat's-blood recipes can be seen as falling along a spectrum on which actual experience and practice stands at one end and the pure articulation of principles stands at the other. For example, in a 1531 collection of recipes that can be found in several other variants entitled the *Rechter Gebrauch der Alchimei*, there are several recipes to make precious stones capable of cutting and of casting. One calls for goose and goat's blood which is to be dried, pounded and mixed with willow ashes, then boiled and mixed with strong vinegar. The stone to be softened should be warmed in the mixture, which allows the practitioner to 'cut or form it as you want'.[32] This recipe might conceivably be effective for engraving or etching on imitation hardstones made of glass, so it is likely to be a mix of experience and an elaboration of the natural principle by which opposites should be combined.

A 1604 recipe collection, entitled 'The Goldsmith's Storehouse', apparently written by an assayer, contains an even more interesting mix of experience and principles with regard to goat's blood: 'some Aucthors doe wryte that the Dyamon cannot be broken, butt with the new warme bludd of a goate, but it is not soe, for Dyamon Cutters have dayly experience to the Contrary, who doe continually use

---

30 Albertus Magnus, *Book of Minerals*, trans. D. Wyckoff (Oxford: Clarendon Press, 1967), p. 70.

31 Robert Goulding makes this point and gives some examples of recipes in 'Deceiving the Senses in the Thirteenth Century: Trickery and Illusion in the *Secretum Philosophorum*', in Charles S.F. Burnett and William F. Ryan (eds), *Magic and the Classical Tradition* (London: Warburg Institute, 2006), pp. 135–62. Eamon, *Secrets*, also makes this point, pp. 34 and 85–7.

32 *Rechter Gebrauch d'Alchimei* (n.p., 1531), fol. IIIv. See Ferguson, *Bibliographical Notes*, part II, p. 42, for information about the variants of this recipe collection.

the powlder of Dyamons'. It goes on to make the point that nothing will cut or polish a diamond except the diamond itself. While modern scholars might seize upon this 'correct' observation that 'must' represent firsthand experience, it is important to notice that the anonymous author of this text has not actually contradicted the goat's-blood recipe, but rather has stated that it is not the *sole* means of cutting a diamond.[33] Indeed, in other sections of this extensive manuscript, the author provides a curious (to our ears) combination of spiritual, material and monetary values inhering in precious and semi-precious stones. For example:

> Coralius or Corallus, is corrall, which is a wede betweene a herbe and a tree, that groweth in the bottome of the sea, smale & braunched like hornes, but softe & whyte so longe as it in the water, it is like a weede or woodde, but taken from the sea, it cometh to the Ayre, it is turned into a stone, – This stone groweth chefely in the sea that is by Massilia, There is two sortes of it, the one reddish like old Ivorye, the other whyte, & is figured in manner like the braunches of trees, the Redder sorte being polyshed, as is first fyled, then polyshed with Sande and after with Tripole & oyle; is most excellent redd but the whyte coullor never taketh any other collour but whyte. It is experienced to prevaile against the fluxe of bludd, and being hanged about the necke it is good against the falling sicknes, and the worke of Menstne which is woemens termes, & against tempests of thunder & lightninge/ The fayer redd coullor being well polyshed, and fairely braunched is worth 2- [?] the oz but unpolyshed not above 2…the oz by reason of the waste & labor in the fyling & polyshinge.[34]

Monetary value is just one of the important active virtues of these powerful stones. The author exhibits a similar mix of values when describing the agate: he notes that because the agate contains the colour red, it preserves eyesight. Moreover, it 'cherisheth the hearte of man[,] being holden in the mowthe, it quencheth thurste, It is good against the stinging of scorpions, and against any other kind of venom or poyson'. But he concludes that the value of agates has gone down, not because the virtue is in any way diminished but because they are more plentiful in his day.[35] Again, the health, medical and monetary virtues of the stone are not regarded as separate. About the magnet, he writes:

> The vertue of this stone is wonderfull, especially for Navigators or Seamen, which cannot sayle with theire Compass without the helpe of this stone. It is sayed that this stone being putt under the hedd of a women sleping by her husband if she be chast, she will presently take her husband and imbrace him, but if she be not so she shal be so troubled with Phantasies & dreames, that she

33 'Goldsmith's Storehouse', fol. 62v.

34 Ibid., fol. 76v.

35 Ibid., fol. 68r.

> shall fall out of the bedd. The Indians doe use to eat daylye of this stone, to make them looke younge.[36]

To a modern person, the mixture of experiential and textual authority from which this entry is derived is troubling, but this statement – as is the case with other discussions of gemstones – demonstrates perfectly the similar authority of both experience and text. This is even clearer in the author's comment on unicorn horn, in which he reports on a textual authority's experimental practices:

> And this Experiment hath bene proved (by my Authore) for vertue of Unicornes horne. That he hath caused two doggs to [be] poysoned, to the one he hath given double quantity of poyson, & to him he gave good quantitie of Unicorn's horne in Powlder as scraped with water, & the dogg hath bene well againe. To the other he gave less quantity of poyson and no unicorn's horne and the dogg presently dyed.[37]

These passages all show clearly these collections of techniques can cover a wide spectrum from firsthand experimentation (with unicorn horn, no less!) to the citation of textual traditions about the virtues of precious stones.

Vernacular Philosophy

As evidenced by the coral and agate entries, red is regarded as indicating the power of these stones to strengthen and heal, not unlike the humour of blood in medical thought. All kinds of red substances and blood appear among the ingredients and operations of recipe collections, not just goat's blood, but ox blood, fox blood, ram's blood, the blood of a menstruating woman, and dragon's blood – about which the anonymous assayer of 'The Goldsmith's Storehouse' refers to Aristotle's description of it as a stone which 'is very redd, it prevayleth against any fluxe of Bludd', but notes that, on the other hand, some physicians say it is the juice of an herb.[38] The colour red also appears to possess many of the attributes of blood in recipes and it too is mentioned frequently in recipes. Let us focus for a moment on this point, for the large number of recipes that contain blood and/or red substances alerts us to the importance of this ingredient.

36 Ibid., fol. 84v.

37 Ibid., fol. 97v.

38 Ibid. On different sorts of blood, see for example the recipes in the collection written down in the Monastery of Tegernsee in the fourteenth and fifteenth centuries: Anna Bartl, Christoph Krekel, Manfred Lautenschalger, Doris Oltrogge, *Der 'Liber illuministarum' aus Kloster Tegernsee* (Stuttgart: Franz Steiner Verlag, 2005), and especially the discussion by Christoph Krekel and Manfred Lautenschlager, 'Bearbeitung von Glas, Edelstein, Bein und Horn', on p. 675.

We can examine what blood might signify in recipes by examining the painter Cennino d'Andrea Cennini's 'book of the art' (*Il libro dell' arte*) written down in Padua at the very end of the fourteenth century. Why did Cennino write down his working practices and his recipes? Cennino's book replicated the complete course of an apprenticeship for a painter, from picking up the chicken bones under the tables and charring them for charcoal to be used in drawing through pigment-making, panel painting, gilding, frescoing and casting. His book records painters' techniques current in the fourteenth and fifteenth centuries and is a good example of a collection of recipes that appears to have been meant for instruction. But even its apparently straightforward recipes yield up an organizing set of principles. Cennino wrote his book, as he tells us, to establish that the art of painting requires both *scienza* (science) and *operazione del mano* (work of the hand). Cennino nowhere defines *scienza* very clearly, but for him it has two important characteristics: first, it is a superior form of activity and lends this superior status to whoever articulates it, and second, it involves the work of imagination and intellect. Cennino was making a claim for his expertise within a hierarchy of knowledge that placed the written word and *scientia* higher than practice and embodied knowledge. His transition from practicing skills and techniques to writing them down in his 'book of the art' – in common with many other compiler/authors at about the same time – was about asserting his identity, his modes of cognition, his skills and his own particular kind of knowledge.

Another aim of Cennino's book was to make clear his religious piety and the spiritually-sanctioned nature of handwork.[39] Cennino begins his text with an abbreviated recounting of Genesis but he moves immediately to the most important part, the Fall. As he says, 'Inasmuch as you have disobeyed the command which God gave you, by your struggles and exertions you shall carry on your lives…'.[40] Cennino's painting and his text were both works of redemption through labour. In one of his recipes, he describes precisely how one is to lay in the flesh tones of living individuals and specifies that this flesh tone is never to be used on dead faces. He called this colour '*incarnazione*' and clearly regarded its use as akin

39 Cennino D'Andrea Cennini, *Il Libro dell'Arte (The Craftsman's Handbook)*, trans. D.V. Thompson, Jr (New York, NY: Dover, 1960). Cennino begins and ends the book with a prayer to 'God All-Highest, Our Lady, Saint John, Saint Luke, the Evangelist and painter, Saint Eustace, Saint Francis, and Saint Anthony of Padua' that the students of his book will 'study well and … retain it well, so that by their labors they may live in peace and keep their families in this world, through grace, and at the end, on high, through glory, *per infinita secula seculorum*', p. 131. On beginning every panel, a painter would invoke 'the name of the Most Holy Trinity … and that of the Glorious Virgin Mary', p. 65. And when the painter begins the extremely delicate task of drawing with a sharp-pointed needle on gilded glass, being careful not to make a single mistake, he must begin 'with the name of God', p. 113.

40 Ibid., p. 1.

to the incarnation of life in a body.[41] This giving of life to (or 'incarnating') an image for Cennino was a straightforward artisanal technique by which the abstract principle and profound miracle of the incarnation of God and the Word in human flesh could be imitated.[42] This simultaneously material and spiritual understanding of the production of materials illustrates the 'theory' that underlay artisanal practices, although it was a lived and practiced theory, rather than a written and abstracted one. It is a commonplace that such piety was a component of the *quattrocento* world, but in Cennino's book, it reveals itself as part of an underlying set of principles that gave meaning and order to the world; it was knowledge that explained 'why', yet it is contained in a 'how-to' book, largely made up of recipes.

It turns out that a further explanatory theory of sorts – a kind of vernacular science – underlies other recipes in Cennino's book. In 'How to Paint Wounds' he instructs the painter 'take straight vermilion; get it laid in wherever you want to do blood'.[43] The pigment vermilion was an artificially manufactured substitute for naturally occurring cinnabar (mercury sulphide) which painters had employed as a deep red pigment in Europe and Asia from ancient times. To manufacture the artificial red pigment, mercury and sulphur were heated together until they become a black paste. On constant heating and stirring, this paste turned dark blue on the outside and silver on the inside, and eventually formed a vapour that condensed as a bright red cake on the walls of the crucible. This powder was then scraped off to form the pigment.[44] The red of vermilion was associated naturally with blood and in particular with the blood of Christ. For example, in indicating colours to be employed in illuminating medieval manuscripts, scribes often used a cross to indicate where the red pigment vermilion was to be used.[45] Perhaps the cross even referred to the making of vermilion in the *crucible*; the root word of 'crucible' in English and Italian is 'cross', and it was in the crucible that sulphur and mercury underwent their own passion and transformation to produce the blood

---

41 Christiane Kruse, 'Fleisch werden – Fleisch malen: Malerei als "incarnazione". Mediale Verfahren des Bildwerdens in Libro dell'Arte von Cennino Cennini', *Zeitschrift für Kunstgeschichte*, 63/1 (2000): pp. 305–25. See also Ann-Sophie Lehmann, 'Fleshing Out the Body: The "Colours of the Naked" in Workshop Practice and Art Theory, 1400–1600', in Ann-Sophie Lehmann and Herman Roodenburg (eds), *Body and Embodiment in Netherlandish Art* (Zwolle: Waanders, 2008), pp. 86–109.

42 Through such a practice, Cennino acknowledged the transformative power of art and artisan.

43 Cennini, *Il libro*, p. 95.

44 Daniel V. Thompson, Jr, 'Artificial Vermilion in the Middle Ages', *Technical Studies in the Field of the Fine Arts*, 2 (1933–4), pp. 62–70.

45 John Gage, 'Colour Words in the High Middle Ages', in Erma Hermens (ed.), *Looking Through Paintings: The Study of Painting Techniques and Materials in Support of Art Historical Research (Leids Kunsthistorisch Jaarboek XI)* (Baarn, The Netherlands: Uitgeverij de Prom, 1998), pp. 35–48 (p. 39).

red pigment. In any case, as we have already noted, blood was regarded as an extremely powerful agent in recipes.

Moreover, the colour red was also associated with gold, first because gold was recommended to be drunk as potable gold or worn on the body in order to strengthen the blood. Second, according to metalworkers' beliefs and alchemical theory, gold was made of a mixture of elemental sulphur and mercury; thus both the red pigment vermilion and gold were composed of mercury and sulphur. Moreover, alchemists believed that base metals could be transformed to noble ones by means of a red powder known as the Philosophers' Stone. Red additives, such as vermilion, were often added to recipes for making gold pigments or in processes involving gold, even when the vermilion was unnecessary to the process.[46] Cennino himself gives a recipe for mosaic gold, a sparkling golden pigment that imitated pure gold, and this recipe included mercury and sulphur, although only the sulphur is actually necessary for the production of this pigment.[47] But these two components, ingredients of both vermilion and gold, seem to have been viewed as necessary to regenerate and transform, and especially to produce gold colour. Cennino's bare-bones recipes thus give insight into a network of beliefs that underpinned his work. Red, vermilion, blood and gold formed nodes in this network.[48] In Cennino's book, the recipes can be followed by an aspiring painter and, at the same time, they reveal the explanatory framework in which this practice took place. Consequently, his book of recipes of his art is straightforwardly didactic, yet it also conveys an explanatory or theoretical framework.

Blood is also involved in the strangest recipe contained in the twelfth-century metalworker Theophilus's *De diversis artibus*. His recipe for Spanish gold called for 'red copper, basilisk powder, human blood, and vinegar'. In order to produce the basilisk powder, two 12 to 15-year-old cocks were put into a cage, walled like a dungeon with stones all around. These cocks were to be well fed until they copulated and laid eggs, at which point toads then should replace the cocks to hatch the eggs, being fed bread throughout their confinement. Male chickens eventually emerged from the eggs, but after seven days they grew serpent tails. They were

---

46 For example, 'Goldsmith's Storehouse', Ch. 25, fol. 55r: 'To make salpeter with Vermilion for water that will part gold from silver as well as aquafortis. 4 oz of vermilion added to the lye, boil etc until it become salpeter…'. Spike Bucklow, 'Paradigms and Pigment Recipes: Vermilion, Synthetic Yellows and the Nature of Egg'. *Zeitschrift für Kunsttechnologie und Konservierung* 13 (1999), pp. 140–49, cites pigment recipes in which red and gold are associated, pp. 145–7.

47 Cennino Cennini lists one such recipe, which calls for 'sal ammoniac, tin, sulphur, quicksilver, in equal parts; except less of the quicksilver'. Cennini, *Il libro*, 'Mosaic gold', pp. 101–2.

48 I say more about this vernacular science of nature in Pamela H. Smith, 'Vermilion, Mercury, Blood, and Lizards: Matter and Meaning in Metalworking', in Ursula Klein and Emma Spary (eds), *Materials and Expertise in Early Modern Europe: Between Market and Laboratory* (Chicago, IL: University of Chicago Press, 2010), pp. 29–49.

to be prevented from burrowing into the floor of their cage by the stones, and to further reduce the possibility of escape, they were to be put into brass vessels 'of great size, perforated all over and with narrow mouths'. These were closed up with copper lids and buried in the ground. The serpent-chickens, or basilisks, fed on the fine soil that fell through the perforations for six months, at which time the vessels were to be uncovered and a fire lit under them to completely burn up the basilisks. Their ashes were finely ground and added to a third part of the dried and ground blood of a red-headed man which was then tempered with sharp vinegar. Red copper was to be repeatedly smeared with this composition, heated until red-hot then quenched in the same mixture until the composition eats through the copper. It thereby 'acquire[d] the weight and colour of gold' and was 'suitable for all kinds of work'.[49]

This recipe has vexed historians, especially because it is the only fantastic sounding recipe in Theophilus's book, a book which otherwise contains eminently clear and practical instructions for carrying out metalworking procedures. Historians have variously viewed this odd recipe as a garbled set of instructions for making brass or for the chemical process of cementation in which gold is purified.[50] Arie Wallert has interpreted it as an alchemical recipe, with blood forming an alchemical 'cover name' (a secret term known only to the *adepti*) for sulphur and basilisk ash for mercury.[51] Sulphur and mercury were the essential components of the philosophers' stone to turn base metals – in this case copper – into gold. In his fascinating study of the relationship between red and yellow pigment recipes, Spike Bucklow views this recipe as evidence for the centrality of the vermilion making process as providing the model for other processes of metallic transformation.[52]

Rather than viewing it as problematic because it seems impossible to carry out, perhaps, with the example of Cennino's use of vermilion in mind, we can view it instead as conveying something other than practice. Theophilus's recipe is remarkably similar to a later set of recipes which call for lizards in place of basilisks. In a 1531 text that includes pigment-making and metalworking recipes, there are several recipes for making noble metals through a process of catching, feeding and burning lizards. As in the instructions for softening hard stones by means of goat's blood, this recipe opens with quite precise instructions on how to catch these lizards. It directs the reader to move very quietly in 'felt slippers', to quickly snatch the lizards before they give off their poison and then to immediately plunge

---

49 Theophilus, *The Divers Arts*, pp. 119–20.

50 Robert Halleux, 'The Reception of Arabic Alchemy in the West', in Roshdi Rashed (ed.), *Encyclopedia of the History of Arabic Science* (3 vols, London and New York, NY: Routledge, 1996), vol. 3, pp. 886–902 (pp. 887–8), argues that it is based on Arabic recipes, perhaps part of the Jabirian corpus.

51 Arie Wallert, 'Alchemy and Medieval Art Technology', in Z.R.W.M. von Martels (ed.), *Alchemy Revisited* (Leiden: Brill, 1990), pp. 154–61 (p. 161).

52 Bucklow, 'Paradigms and Pigment Recipes', p. 145.

them into a pot of human blood. A recipe for making 'lizard-rib gold' follows, which calls for two pounds of filed brass and a quart of goat's milk, and continues: In a pot wide at the bottom and narrow at the top, with a cover that has airholes in it, place nine lizards in the milk, put the cover on, and bury it in damp earth. Make sure the lizards have air so that they do not die. Let it stand until the seventh day in the afternoon. The lizards will have eaten the brass from hunger, and their strong poison will have compelled the brass to 'transform itself to gold'. Heat the pot at a low enough temperature to burn the lizards to ash but not to melt the brass. Cool the mixture, then pour the brass into a vessel, rinse it with water, then put it in a linen cloth and hang it in the smoke of *sal ammoniac*. Once it is washed and dried again, it will yield a 'good calx solis', or powdered form of gold.[53]

This recipe may employ lizard ash as a cover name for mercury, or the recipe may be to produce the painter's pigment mosaic gold. What is important, however, is that in this and other lizard recipes in this collection, lizards are associated repeatedly with generation and transformation, in part apparently through their connection with mercury and sulphur, the generators of gold and blood-like vermilion. Another trace of this association of lizards with metals and transformation is found in a book of secrets ascribed to Albertus Magnus, written no later than the fourteenth century. Among many secrets for lighting a house, one calls for cutting off the tail of a lizard and collecting the liquid that bleeds from it, 'for it is like Quicksilver', and when it is put on a wick in a new lamp 'the house shall seem bright and white, or gilded with silver'.[54] Lizards were believed to spontaneously generate out of putrefying matter, and they regenerated their tails when they lost them. Furthermore, they emerged fully grown from the ground after freezing winters. In other words, lizards were creatures that could give insight into processes of putrefaction, generation, and regeneration. It is thus not surprising that they were associated with transformation in the recipe collections.

Attitudes to Matter

As in the recipes involving blood, it seems more fruitful to view these instructions involving lizards in the *Kunstbücher* as setting out the theoretical underpinnings of a craft rather than practical directions for actual operations. Thus, the lizard recipes seem to be another node, along with blood, red and gold, in a theoretical framework by which metalworkers explained processes of generation and transformation. But what about the naturalistic details of these recipes – for example, those about approaching the lizards carefully and keeping them from suffocating by perforating their vessels? Perhaps these should be understood as inculcating certain habits necessary to craft practice. The wearing of felt slippers and the concentration of the lizard hunt, for example, proclaim 'stay alert!' One of

53 *Rechter Gebrauch*, fol. XIIIr.

54 Albertus Magnus, *Book of Secrets*, p. 104.

the most important components of successful handwork is the training of attention, and many recipes make reference to this important habit of being wide awake to the changes in the state of the material and the task at hand.

The injunction not to suffocate the lizards seems to say to the practitioner: put yourself in the position of the material, both enjoining the practitioner to know the materials with the whole body, and also perhaps alluding to the mystical goal of union with the material in order to gain knowledge.[55] The burning of lizards to a temperature at which the lizards have turned to ash but brass is not yet melted might indicate the importance of judging temperature and knowing the behaviour of materials on heating, for, as Vannoccio Biringuccio, the author of the 1540 metalworking text *Pirotechnia*, noted about metalworking: the regulation of fire is 'a very important thing at the beginning, middle, and end, because it is the principal agent'.[56]

These instructions that seem to ask the reader to come to 'feel', to 'sense' and to 'know' matter resemble Cennino's deceptively homely discussion of materials when he speaks about the wooden panel on which he is going to paint as being hungry – and having to give it an appetizer of size (a thin gesso mixture) before laying on the following coats which constituted its meal. Or, as Biringuccio in 1540 said of the lodestone: 'it appears that Nature has put into this stone a certain spirit of vivacity, so that it seems to have – I would like to say – hands, although they are not seen'.[57] Matter was like a living being that one had to come to know through intimate and bodily acquaintance. It would appear that these recipes attempted to convey the attitudes necessary to the craft, an important component of the 'mystery' or 'secret' of the craft that could also be reinforced by initiation procedures.

Recipe collections, then, could encourage an attentive meditation on materials; they could develop a habit of regarding matter and its manipulation. Moreover, by their very repetition, often listing different variations of ingredients or different methods of doing an operation – sometimes simply titling them 'another way' – recipes could encourage trial-and-error testing. Such trial-and-error procedures gave the 10,000 hours of practice in techniques that cognitive psychologists say are needed to gain expertise, but, more than this, the repetition with variation teaches that matter is something to work through, and something through which to explore resistances, and in which to seek out the characteristics of a material in different situations.[58] The need for repeated trial is emphasized by Biringuccio

55 Union with the material through 'experiment', or 'overhearing', as Paracelsus called it, was the means by which artisans gained knowledge of the powers or virtues inherent in matter.

56 Biringuccio, *Pirotechnia*, p. 198.

57 Ibid., p. 114.

58 A long collection of recipes and techniques written down by a French practitioner at the end of the sixteenth century is overwhelmingly about exploring the resistances of materials. On BnF, Ms. Fr. 640, see Pamela H. Smith and Tonny Beentjes, 'Nature and

in the *Pirotechnia*: 'those things that have such inner powers [implanted by God], like herbs, fruits, roots, animals, precious stones, metals, or other stones, can be understood only through oft-repeated experience'.[59]

Such a grounding in the habits of practice and the behaviour of matter led to an ability to intuit, improvise and innovate. As Biringuccio put it: the practitioner must 'understand well what you wish to do in this operation, and for you to adapt the force to overcome the resistance that is offered according to the qualities of the materials ... But because the light of judgement cannot come without practice, which is the preceptress of the arts, I shall pass through this briefly with the idea of one day being able to supplement this further by demonstrating it to you'.[60] Judgement could only come through practice, not through words or texts, and a physical demonstration really was worth a thousand words.

The hours of practice enabled the practitioner to respond to the unknown; this was a training in coming to be capable of intuition, a repetition of particular instances and experiences until they become generalized as 'second nature'. Improvisation based on thorough knowledge is the stock in trade of the practitioner: In the smoke and heat of the workshop, with dangerous and molten materials all around him, the metal caster must make the split second determination at which exact moment the metal is ready to pour. Technical writing – books of secrets – sought to convey the necessity of educating the attention, of overhearing or thinking with matter, and of repetitive 'doing' in order to respond to the contingencies of the workshop. As a modern Japanese silversmith trained in traditional methods put it:

> Remember, our work is not done by measuring and talking. The hammering, the forging, all the processes are performed by intuition. It's the split-second intuitive decision to remove the iron from the fire, when and how to bring up the flame, to immerse the blade in the water now – it is these acts of intuition that produce.[61]

And, as the 1604 assay master of 'The Goldsmith's Storehouse' phrased it:

> a pfitt Assay Master [is one] whose perfection [is] grounded upon Artificiall Exercise, for these thinges doe rather consiste in doing then in referringe, for

Art, Making and Knowing: Reconstructing Sixteenth-Century Life Casting Techniques', *Renaissance Quarterly*, 63/1 (2010): pp. 128–79. I found Richard Sennett, *The Craftsman* (New Haven, CT and London: Yale University Press, 2008), very useful in thinking about the function of recipes.

59 Biringuccio, *Pirotechnia*, p. 114.

60 Ibid., p. 280.

61 Suzanne B. Butters, *The Triumph of Vulcan: Sculptor's Tools, Porphyry, and the Prince in Ducal Florence* (2 vols, Florence: Olschki, 1996), vol. 1, pp. 286–7, quoting from E. Lucie-Smith, *The Story of Craft: The Craftsman's Role in Society* (Oxford: Phaidon, 1981), p. 85.

> they are not easelye reduced to matter of Argument … the trade asketh a good Judgment, gotten rather by years & experience, then by speculation & dispute.

The assay master must possess:

> grounded experience in this Science or mysterie, should have a perfect Eye to vewe, & a stedye hand to waye for other mens senses cannot serve him.[62]

Conclusion

What then is a secret or a recipe? It could be a straightforward set of instructions and ingredients, but it might also lay out the 'vernacular science' of the author – the principles by which he oriented himself – for example, as has been laid out briefly above, the four elements and humours, the power of the heavens over earth, or the nexus of blood, red, gold, and lizards. But another, perhaps less obvious goal of the collections was to convey that the 'secret' of craft lay in the ability to attend closely, to know the characteristics and behaviour of materials, and to respond to new situations intuitively. Like Cennino's book of the art, they sought to convey what was learned in the 10,000 hours of an apprenticeship. Books of secrets, recipes, and techniques, then, sought to make clear that the ability to intuit and improvise was an extremely powerful combination of practice and thought. We might today consider it an embodied equivalent to the functioning of generalization in induction. In the epistemology of practical writing, intuition was the certain *scientia*, improvisation was the generalizing action and the proof was in the success attained by producing the right result or making things work. As such, these practical 'how-to' books included thinking about doing and how to train for it, but they also included thinking about thinking, if that includes such elements of cognition as attentive action and intuition. In other words, books of secrets recorded ingredients and operations, but they also attempted to convey the secret of embodied cognition.

[62] 'Goldsmith's Storehouse', fols. 5v–6v.

PART II
Secrecy and Openness

Chapter 3
The Secrets of Sir Hugh Platt

Ayesha Mukherjee

In 1594, the Elizabethan scientist Hugh Platt (1552–1608) published his best-known book *The Jewell House of Art and Nature*, which he prefaced with a critique of the 'professors of secrets'. Platt provided a list of these authors whom he, with a disparaging edge, called 'that Magical crew': 'Albertus Magnus, Alexis of Piemont, Cardanus, Mizaldus, Baptista Porta, Firouanta … Wickerus'.[1] Many of these authors, as William Eamon has shown, were reputed to have travelled widely, gathering secrets from varied sources; some founded private academies for the purpose of exclusive intellectual exchange and cooperation, and some expressed disenchantment with established approaches to knowledge.[2] The initial appeal of such figures to Platt, a Cambridge-educated gentleman of Lincoln's Inn who delighted in his own intellectual virtuosity, is easily understood. But his critique of popular works by professors of secrets is a complex issue informed by social and cultural change in late Elizabethan England.

This chapter will examine the intricate ways in which Platt expressed and justified his criticism of books of secrets and of secrecy itself. The 'secrets' of Sir Hugh Platt were designed to be revealed, widely disseminated, and modified by practice. I will discuss his actual utilization and economic application of books of secrets and analyze the distinctive rhetorical and performative stances which guided his modified publication of secrets drawn from the texts at his disposal. My examination of Platt's critique will thus combine the evidence of his published works, his manuscripts, and indeed his marginalia in his own copies of books by some of the authors he lists. My aim is to closely read the reading of books

---

1 Hugh Platt, *The Jewell House of Art and Nature* (London, 1594), sig. B3v. The authors of 'books of secrets' named here are: Albertus Magnus (c. 1206–80), 'Alessio Piemontese' (fl.1514–47), Girolamo Cardano (1501–76), Antoine Mizauld (1510–78), Giambattista della Porta (c. 1535–1615), Leonardo Fioravanti (1518–88) and Johann Jacob Wecker (1528–86).

2 A well-known example of such an academy is the *Accademia Segreta* founded by Girolamo Ruscelli (c. 1500–1565/6) who claimed to be the real author of the *Secrets* of 'Alessio Piemontese'. Ruscelli's academy was a secret society which, he claimed, would eventually publicize its work, but only after its experiments had been 'reduc[ed] to certainty and true knowledge'. See William Eamon, *Science and the Secrets of Nature: Books of Secrets in Medieval and Early Modern Culture* (Princeton, NJ: Princeton University Press, 1994), esp. pp. 147–51.

of secrets by an Elizabethan figure who was, with good reason, well known in his time both as an author and as an expert practitioner of natural knowledge.[3] The close reading of reading, as the recent work of William Sherman and others has shown, is itself a complex task: it requires sensitivity towards the varied and sophisticated notions of both 'use' and 'reading' in a cultural context.[4] Indeed, such is the complexity and sophistication of Platt's reading that there will not be space in this chapter to consider all the books I have so far discovered from his library or even all of his marginalia in his copies of books of secrets in particular. But I will attempt, through focused attention to specific uses of selected books and the contexts of their use, to illustrate Platt's methodology and economic application of the literature of secrets available to him.[5]

Platt and the Professors of Secrets

Platt's criticism of the professors of secrets is scattered across several published works as well as drafted prefaces and receipts among his manuscripts.[6] His objections are based on the tension between the notion of secrecy and pragmatic need. I will explore this by first examining his disapproval of the language used by the authors of secrets and the formulation of Platt's own rhetorical strategy. He complained that not only did these authors write in foreign languages, thereby making English translations a necessity, they also employed forms of deliberate

---

3 Although Platt was well known in his own times and his work was regarded with respect by the early members of the Royal Society, there are few modern scholarly accounts of him. Brief, but useful, descriptions of some of Platt's work are found in Eamon, *Science*, pp. 311–14, and in Charles Webster, *The Great Instauration: Science, Medicine and Reform, 1626–1660* (London: Gerald Duckworth and Co., 1975), pp. 468–70. Also useful is Deborah Harkness's comparison of Platt and Bacon in *The Jewel House: Elizabethan London and the Scientific Revolution* (New Haven, CT: Yale University Press, 2007), pp. 211–53. A full account of Platt's context, reading and works can be found in my doctoral thesis: Ayesha Mukherjee, *Food and Dearth in Early Modern England: The Writings of Hugh Platt* (PhD thesis, University of Cambridge, 2007).

4 See esp. William H. Sherman, *Used Books: Marking Readers in Renaissance England* (Philadelphia, PA: University of Pennsylvania Press, 2007); the seminal work of Roger Stoddard, *Marks in Books, Illustrated and Explained* (Cambridge, MA: Harvard University Press, 1985); and Bradin Cormack and Carla Mazzio, *Book Use, Book Theory, 1500–1700* (Chicago, IL: University of Chicago Press, 2005).

5 See Sherman, *Used Books*, p. xvi, on the importance of balancing the narratives of individual 'mysteries' in Renaissance used books with broader accounts of more 'large-scale patterns of use'.

6 See, for example, 'To the studious and well-affected Reader' in Hugh Platt, *Floraes Paradise* (London, 1608), sig. A4v; Hugh Platt, *The New and Admirable Arte of Setting of Corne* (London, 1600), sig. C1v; Hugh Platt, British Library (hereafter BL), Sloane, MS 2197, fols 1–4. These examples, and others, are further discussed below.

obfuscation ('obscure and aenigmaticall' phrasing) which especially irritated Platt.[7] The philosopher Cornelius Agrippa, Platt protested, wrote learnedly, 'though exceeding darklie', in the three books of his *De occulta philosophia* (1509–10). Platt had translated these into English and intended to publish his translation.[8] Similarly, the much-admired Italian scholar della Porta penned aspects of 'vegetable Philosophy' 'as a Sphinx, and rolled it vp in the most cloudy & darksome speech that he could possibly haue devised'.[9] In opposition, Platt announced his own intention of putting down receipts in the 'plainest and most familiar phrase' that he could devise.[10]

Indeed, his strategy is the precise reversal of penning as a Sphinx and rolling up in clouds: Platt devised instead a rhetorical and performative stance of revelation through rational elucidation. This is captured, for instance, in the descriptions of unusual local phenomena and 'magical' occurrences in his gardening treatise *Floraes Paradise* (1608). Here, the extraordinary fruitfulness or unseasonal blossoming of locally famous trees is explained in terms of the specific chemical and physical environment of plant growth. Platt seems especially excited with a story about 'that delicate knight, Sir Francis Carew', who, during a visit from the Queen, led her to a cherry tree. The fruit of this tree had been kept back from ripening at least a month after 'all Cherries had taken their farewell of England'. This secret, explained Platt, was performed by straining a canvas cover over the tree and 'wetting the same now and then with a scoope or horne, as the heat of the weather required'. So, by 'with-holding the sunne beames from reflecting vppon the berries, they grew both great, & were very long before they had gotten their perfect cherrie-colour'.[11] When he was sure of the Queen's arrival, Carew

[7] Preface in Platt, *Jewell House*, sig. B3v. The texts of Platt's own published works often contain his embedded translations of extracts from other authors: *Arte of Setting Corne* (1600) contains Platt's translations of passages from della Porta (sig. C1v); and *Diuerse New Sorts of Soyle* published with a separate title page in *Jewell House* (1594), contains summaries (and often exact translations of extracts) from Bernard Palissy's *Discours admirables* (1580) and Francisco Vallés' *De sacra philosophia* (1592) followed by Platt's elucidation of the passages (sigs A2r-D2v).

[8] Preface in Platt, *Floraes Paradise*, sig. A5r. The translation or 'Abstracte' was probably not published but the manuscript version can be found in BL, Sloane, MS 2223, fols 30–47: 'An Abstracte or Epitome of a large & theoricall Discoorse vppon the Animal and vegetable stone, gathered owte of the three bookes de occulta Philosophia, written by y[t] greate Scholler Corn. Agrippa'. As the title shows, apart from translating, the very process of summarizing and selective gathering was itself a means of explication which Platt considered important.

[9] Preface in Platt, *Floraes Paradise*, sig. A5r.

[10] Ibid.

[11] Platt, *Floraes Paradise*, pp. 174–5. Platt's account is quoted by John Nichols in *Progresses and Entertainments for Elizabeth* (London, 1823), vol. 3, p. 441. Carew's Beddington home was frequently visited by the Queen who, to the courtier's alarm, grew to expect 'greater entertainment and gifts' from him. See Felicity Heal, 'Giving and Receiving

uncovered the tree and a few sunny days brought its fruit to full maturity. Platt's reader is left to imagine the Queen's unexpected pleasure when she beheld this miraculous plenty at a time when there was meant to be a dearth of cherries in the country.

The pattern of concealment and revelation in the narrative of unveiling the cherry tree's untimely produce reminds us of the provocation of wonder through studied nonchalance or *sprezzatura*, which Eamon identifies in the writings of the Italian *professori secreti* and aptly allies with their courtly self-fashioning and ambitions.[12] While the nuances of this attitude are expressed in Platt's story about Carew and the Queen, the voice of the author, who determinedly narrates in 'plain style', is consciously separated from the flourish of the courtier. Platt's voice acquires power not (as the courtier does) from the pleasure induced by the miraculous display itself, but from the dissemination of the display's secret to readers who are urged to be active in making their own trials of such receipts. Platt thus constructs his authorial voice and rhetorical strategy in resistance to della Porta's dictum, 'If you would have your work appear more wonderful, you must *not* let the cause be known'.[13] Unlike professors of secrets such as della Porta or Ruscelli, who resisted revelation, especially of unperfected secrets, Platt sought to involve the reader in the very process of perfecting knowledge.

In Platt's rhetoric, this desire to reveal develops through associations between secrecy and starvation, on the one hand, and between cooperative endeavour and sufficiency, on the other hand. He frequently describes disseminating knowledge as a process of feeding malnourished stomachs with 'secret foode' or as the provision of pragmatic aids for the economically deprived. In a treatise called *A Discovery of Certaine English Wants*, published in 1595, Platt called for the systematic sharing of knowledge which could lead to the taking of cooperative and congruous measures across different economic units. Such sharing, he claimed, would be to the aid of any 'poore Lazarus, who neither as yet with the saltish drops of his body, is able by his labour, nor yet by the bleeding tears of his soule is able by his misery to procure sufficient maintenaunce the one way, or compassion the other way, to releiue himself and his family'.[14] An especially poignant aspect of this treatise is the way it shapes Platt's economic and moral vocabulary: it repeatedly relies on the balanced ironies created by Platt's modulation of the notions of 'wanting', 'spending', 'saving', 'profit', 'sufficiency', 'maintenance',

on Royal Progress', in Jayne Elisabeth Archer, Elizabeth Goldring, and Sarah Knight (eds), *The Progresses, Pageants, and Entertainments of Queen Elizabeth I* (Oxford: Oxford University Press, 2007), p. 59.

12 As Eamon observes, the art of the professor of secrets 'resembled the masterful masquerading of the courtier, whose "artful dissimulation" won the favor of a prince or lady without flattery or force'. See Eamon, *Science*, p. 226.

13 Giambattista della Porta, *Natural Magick*, English translation (London, 1658), p. 4.

14 Hugh Platt, *A Discovery of Certaine English Wants* (London, 1595), sig. A3r. A manuscript draft of this treatise exists in BL, Sloane, MS 2197, fols 1–4.

'substance', 'moiety', 'charge'. All these words become inextricably allied to the production of 'substantial' and 'creditable' experiments that are 'new, & fresh from the mint'. This alliance is built up to support Platt's explicit suggestion that the work of 'good wittes' required additional tests of its economic effects before one could admit that this work was productive of 'publike good'.[15] In this light, one may review Platt's seemingly conventional claim that the receipts of della Porta and others had to be subjected to hands-on trial 'in the glowing forge of Vulcan', and those receipts which proved 'substantial' sifted from those which were likely to 'vanish into smoake'.[16] A close analysis of Platt's language thus demonstrates his economic and utilitarian grounding of the role of knowledge and its revelation. I will argue that this point of view was influenced by Platt's own socio-economic position, which shaped the ways in which he read and utilized books of secrets, and it is therefore important to reconstruct the local environments in which such books were read.

Platt's Milieu and Networks

Hugh Platt was the son and main heir of a wealthy London brewer, Richard Platt, who served as Master of the Brewers' Company and Sheriff of London, endowed a free grammar school and almshouses in his native Hertfordshire, and held various properties in London and its suburbs. Richard managed a chain of brew houses which also functioned as retail outlets and were subsequently supervised by Hugh and his sons.[17] Hugh Platt graduated BA from St John's College, Cambridge, and was later admitted to Lincoln's Inn. He published, at the age of 20, a work called *The floures of philosophie*, which was probably a version of his commonplace book. This included culled precepts from Seneca and a set of poems by Platt that sharply satirized contemporary socio-economic practices such as the racking of rents, enclosures, bad husbandry and covetousness. A long alphabetical index to this work categorized the Senecan precepts under broad topics that connect with the moral themes of Platt's own poems. This indexing methodology, inviting readers to engage, sift and construct their own readings, is further refined in indices to Platt's later publications that drew on books of secrets.[18]

---

15 Platt, *English Wants*, sig. A4v.

16 Platt, *Jewell House*, sig. B4r.

17 Biographical information about Richard and Hugh Platt, beyond what is available in the *Oxford DNB*, may be found in the wills of members of the Platt family, records of the Brewers' Company and the Aldenham School in Hertfordshire, and the Platt family papers in the archives of St John's College, Cambridge. For a detailed analysis of these documents, see Mukherjee, 'Dearth Science, 1580–1608: The Writings of Hugh Platt', Occasional Paper 35, *The Durham Thomas Harriot Seminar* (Durham, 2007).

18 Hugh Platt, *Floures of Philosophie* (London, 1572), sigs A5v–A8r. Here, Platt is both following and modifying the familiar Renaissance practice of collecting excerpts of

From the 1580s onwards, Platt's activities (as well as his reading) demonstrate his extraordinary versatility. He oversaw his family's substantial estates and trade, and some of these became the sites of his own agricultural and other scientific experiments which were designed to alleviate the increasing pressures on landowners and traders in an environment driven by economic shortages, overcrowding, threats of war and outbreaks of disease.[19] A significant proportion of the Platt family properties lay beyond the city wall, in the parishes of St Sepulchre and St Pancras. The area, around 1550, comprised much green space: open pasture grounds, fields, orchards and scattered houses. But, by the mid-seventeenth century, these spaces consisted of closely packed tenements with garden plots.[20] As a landowner, Platt monitored such properties in semi-rural suburbs at a moment when these areas were beginning to become developed and populated.[21]

As a student of alchemy, Platt read widely on the subject. Consulting figures such as Godfrey and James Mosanus, John Dee and Moritz of Hessen-Kassel, he searched for an ultimate material source of growth, fertility and nourishment in nature that could be extracted, preserved and reincorporated into products as varied as the farmer's manure, the limner's coloured inks, the housewife's comfits and jams, or Platt's own anti-plague cakes.[22] As an unlicensed medical practitioner,

---

authoritative texts and varied compendia. See Ann Moss, *Printed Commonplace Books and the Structuring of Renaissance Thought* (Oxford: Oxford University Press, 1996); and Ann Blair, 'Humanist Methods in Natural Philosophy: The Commonplace Book', *Journal of the History of Ideas*, 53/4 (1992): pp. 541–51.

19 See esp. John Walter, 'The Social Economy of Dearth in Early Modern England', in Roger Schofield and John Walter (eds), *Famine, Disease, and the Social Order in Early Modern Society* (Cambridge: Cambridge University Press, 1989), pp. 75–128.

20 Compare, for instance, *Woodcut Map of London c. 1550* and William Morgan's *Survey of the City of London* (London, 1682). For details of the St Sepulchre and St Pancras properties belonging to the Platts, see MSS D111-113 in St John's College Archives, Cambridge.

21 Scholarship on the changing face of early modern London is particularly rich: for an overview, see Jeremy Boulton, 'London 1540–1700', in Peter Clark (ed.), *The Cambridge Urban History of Britain 1540–1700*, vol. 2 (Cambridge: Cambridge University Press, 2000). On London's demography and topography, see esp. Roger Finlay, *Population and Metropolis: The Demography of London 1580–1650* (Cambridge: Cambridge University Press, 1981), pp. 51–66; Roger Finlay and Beatrice Shearer, 'Population Growth and Suburban Expansion', in A.L. Beier and Roger Finlay (eds), *London 1500–1700: The Making of the Metropolis* (London and New York, NY: Longman, 1986), pp. 37–59. Also see essays by Vanessa Harding, 'City, Capital, and Metropolis: the Changing Shape of Seventeenth-century London' and Laura Williams, '"To Recreate and Refresh their Dulled Spirites in the Sweet and Wholesome Ayre": Green Space and the Growth of the City', in J.F. Merritt (ed.), *Imagining Early Modern London: Perceptions and Portrayals of the City from Stow to Strype, 1598–1720* (Cambridge: Cambridge University Press, 2001).

22 See Platt's letter to Francis Segar: BL, Sloane, MS 2172, fol.18 (Platt desired to have as intermediaries between himself and Moritz either Segar or 'doctor Mosanus whose

Platt travelled all over the city and beyond, dispensing remedies to the 'LLs of the Counsell', as well as to local servants, millers, grocers, flax wives, lace makers, purse makers, victuallers, vintners, brewers, herb wives, apothecaries, cooks, bakers, painters, joiners, scriveners, goldsmiths, merchants, mariners, coalmakers, gun makers, diamond cutters, tailors, linen drapers, and his own immediate and extended family. He willingly cured poor widows free of charge, accepted the humble payment of two pullets from a Kent coalmaker, and quarrelled with a relatively well-off mercer's wife for not paying so much as his boat hire to Battersea.[23]

As a trader, Platt supervised his father's breweries in various parts of the city, and inhabited the bustling world of commercial gardening in Bethnal Green, living first in Bishop's Hall and later in Kirby Castle, where he maintained and developed the well-known vegetable and flower gardens. These large houses were shared residences which were, like the area of Bethnal Green itself, home to middle-class merchants, professionals and courtiers.[24] Many of these people were Platt's clients and collaborators with whom he exchanged expertise (and books and manuscripts) and evaluated existing knowledge on matters as diverse as alchemy, agriculture, horticulture, husbandry, household management, cookery, as well as the manufacturing and marketing practices of numerous contemporary trades and professions.

The Jewell House

Somewhere in or close to Bethnal Green (the precise location is nowhere mentioned) Platt leased a warehouse and kept a shop known to his acquaintances, clients and associates by the name of 'Jewell House'.[25] This 'Jewell House' stocked items produced by Platt such as 'chaffing dishes to heate without fier', 'reuerberators to roust with', '5 spittes in one', 'Boxes to carrie past or other

father did first teach me to close a helme and a bodye together'); and Platt's notebook: BL, Sloane, MS 2210, fols 26–8, where he describes his meeting with Dee. Some of Platt's reflections on the sources of fertility and growth appear in *Floraes Paradise*, pp. 6–9, and BL, Sloane, MS 2223, fols 30v–1r.

[23] Platt kept records of his medical practice during the years 1593–1605 and 1607–8. See BL, Sloane, MS 2209, fols 16–33 and MS 2203, fol. 113.

[24] T.F.T. Baker (ed.), *A History of the County of Middlesex* (London: Boydell and Brewer, 1998), vol.11, pp. 87–119 and 168–90; *London Metropolitan Archives*, M/093/158, fols 2–4,22 and M/093/142.

[25] See Platt's nearly identical shop inventories titled 'Wares for my new shopp' and 'Wares for my Jewelhowse' (many items on this list have prices attached) in BL, Sloane MS 2197, fols 14–5; and a letter to Platt from his cousin and professional associate Henry Davenport (dated 19 December, 1595) providing further ideas for the 'furnishing' of 'our Jewell House' in BL, Sloane MS 2172, fol. 27.

meates hotte in', 'A waggon for the seamen to draw theyr victualls on land', 'A Lanthorne wherein a naked candle is not blowne owte', 'Yncks powder', 'Paper prepared for blew and redd letters', 'Desks of glas', 'old Tables in oyle refreshed', 'oyle for the Tooth ach', 'collors for hayre', 'cinnamon water', 'glasses to mount claret wyne', 'macaroni and cuscusow [couscous] for the poore', 'oranges, lemons and other fruites owte of season' and numerous other products which carried, as Platt claimed, a unique status.

In other words, the wares sold from 'Jewell House' consisted of items whose receipts were published in *The Jewell House of Art and Nature* and other works by Platt, as well as items whose unpublished receipts may be found in Platt's extant manuscripts. Ideas for many of these receipts were initially conceived via Platt's reading of books of secrets in his library, and subsequently modified by his experimentation and by the demands of Platt's local environment in which the newly evolved 'secrets' were to be published and placed as tradable and usable commodities. These commodities could be the receipt books themselves, which sold well, as Platt was a successful author and something of a publishing phenomenon in his time. They could also be individual shop items listed in such inventories as I have described above. Platt was, additionally, a successful shopkeeper. His commodities could even constitute 'remedies' which might heal not only individual human bodies but also the rapidly changing, frequently crisis-driven national economy. Platt was a much-called-upon medical practitioner in his time and a tireless campaigner for the production and organization of knowledge for civic and economic benefit, or 'public good' as he put it. Consequently, the 'jewels' or 'secrets' Platt found among authors like della Porta or Wecker were, to him and his colleagues, imperfect stones – too cloudy, untested, liable to vanish into smoke – not only because they represented unperfected natural knowledge; they needed sifting, cutting and polishing according to the economic, ethical, as well as aesthetic demands of Platt's specific cultural environment before they could be included in any contemporary 'Jewell House'.[26]

Platt's naming of his shop and his book was, then, no mere metaphorical exercise. The existence of Platt's shop, as well as his constant ethical analysis of and involvement in contemporary markets, makes the use of 'jewel' both more literal and more ironic than we might otherwise assume. Platt was not simply fabricating samples of natural knowledge which he deemed 'jewel-*like*' in their value; their value lay also in their actual manipulative influence on the early modern economy as products and material objects to which prices could be attached. And this in turn would influence the very notions of how nature could be manipulated and how natural knowledge itself could (or should) be shaped and organized.

---

[26] See Harkness, *Jewel House*, where the 'jewel' metaphor is explicated in terms of Platt's collection and improvement of 'unpolished jewels of natural knowledge' (pp. 221 and 224–5). I think, for the reasons given below, that this explains the 'metaphor' only in part. My analysis, by acknowledging the material reality of Platt's shop and products, questions the purely metaphorical status of 'jewel' in Platt's context.

Texts and Contexts

Bearing this in mind, the rest of this chapter will elaborate the colourful and multifaceted context and methodology against which Platt's reading of books of secrets by 'that Magical crew' might, in turn, be read. I thus draw attention to particular books from Hugh Platt's personal library. Part of this library has been fortuitously preserved among the rare books collection of St John's College Library in Cambridge, which contains more than 70 volumes from the libraries of Hugh and his son William Platt. Of these, I have identified 19 works containing Hugh Platt's annotations.[27] Especially relevant to the subject of secrets are three heavily annotated books in this collection: Giambattista della Porta's *Magia naturalis* (1584), Johann Jacob Wecker's *De secretis* (1588), and William Warde's three-volume translation of Alessio Piemontese's *The Secrets of the Reuerend Maister Alexis of Piemont* (1562, 1563 and 1566). In these, Platt's notes are more informal and personal. Annotations were made probably at different times in the course of multiple readings. In the three-volume *Secrets of Alexis*, for example, nearly every secret is marked with these two distinctive symbols:

 and 

The symbols appear to rank the receipts according to their importance in the eyes of Platt. Receipts are marked with either one or both of these symbols, used several times, in various formats, and often in combination. It seems to me that, for Platt, these two symbols were more than mere marks of selection, which abound in early modern used books. The different formatting and clustering of symbols for different receipts suggest an elaborate private coding system showing the number of times the receipts were put to trial.[28] The same marking system is used not only in the della Porta and Wecker volumes but also in Platt's own handwritten notebooks of receipts and experiments.[29] The books contain heavy underlining,

---

27 St John's College Archive, MS U2, fol. 53 and MS U3, fol. 47: these are seventeenth-century lists of books left to the college by William Platt. Not all the listed books are now in the library's possession, but the library has several additional works (not mentioned in the manuscript lists) which were clearly part of the same benefaction. For a list of the 19 works annotated by Hugh Platt, see Mukherjee, 'Dearth Science', p. 27.

28 Typical examples are:

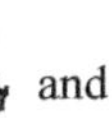 and 

in Platt's copy of *Secretes of Alexis*, trans. William Warde, 3 vols. (London, 1562, 1563, 1566), Part 1, ff.62v and 96v. Held in the library of St John's College, Cambridge; Classmark: Mm. 12.25(1)

29 In fact, one of Platt's favourite symbols  appears enlarged on the titlepage of a notebook – BL, Sloane MS 2216. It is drawn in Platt's homemade copper ink. The entries recorded in the notebook itself are frequently marked with clusters of this symbol inscribed

and marginal notes include cross-references, queries and reminders. In one receipt, for example, which describes how to make the oil of Ben (a popular perfume base made from the seeds of the drumstick tree or the *Moringa oleifera*), readers are instructed to place over a fire various ingredients in 'a vessell such as perfumours vse'. Platt underlines the phrase and writes in the margin 'qre of ye vessell. /' – that is, he queries the precise nature of the vessel. In another case, he marks a receipt for making oil of nutmeg whose printed version recommended soaking the nutmeg in malmsey before crushing it. Platt, attempting to simplify the procedure, writes in the margin 'qre of leavinge owte the malmsey'. Instructions for making 'Bengewyne' (benzoin) oil by sealing the resin in a glass vessel and burying it in a dunghill, are followed by the query 'what other oyles may be made in this way'.[30]

In other words, the text is read and assessed, any vagueness is questioned and Platt's own ideas, variants, or possible alterations are noted. It is clear that receipts in such books were, in Platt's time and environment, being tried in workshops and stillrooms: against a receipt to make a plaster for wounds, Platt notes, 'this my Vncle Birtles comended greatly vnto mee vppon his owne experience/'.[31] The query about the perfumers' vessel is 'answered', in a sense, by the actual designing of such a vessel illustrated in *The Jewell House of Art and Nature*.[32] Similarly, manuscript and printed records of Platt's experiments with various oils – the subject of the two other notes mentioned – are prolific. In short, from the evidence of Platt's library, it is possible to reconstruct the dialogue between his reading and his own writing and experimentation. And through such reconstruction one may track the process by which a secret that appeared obscure and speculative could become accessible, tangible and economically applicable in early modern society. I will now demonstrate this by focusing on a specific area of Platt's work.

'Penury into Plenty'

An outstanding area of Platt's experimentation was the cultivation of grain and its efficient employment in his economy. I will use this as a case study through which we may trace the movement from Platt's reading of books of secrets to their pragmatic application. The trade in grains had a special status in Platt's England. 'Corn', broadly designating the seed of important cereals such as wheat, rye or barley, was a coveted source of staple food. Using alternatives to wheat

in formats similar to the examples found in Platt's copies of the books described here. This suggests, in my view, that Platt subjected his own notebooks to a similar form of active reading that he applied to books in his library.

30 Platt's copy of *Secrets of Alexis*, Part 1, fols 44v, 46v and 51r. Similar examples abound in his copies of della Porta and Wecker (in St John's College library, Cambridge; classmarks: Ee.8.1 and Ll.13.17) and are too numerous to list here.

31 Ibid., fol. 34r. Birtles was the maiden name of Platt's mother, Alice.

32 Platt, *Jewell House*, p. 25.

was common in dearth years, and the husbanding of corn was a crucial issue in agriculture.[33] Platt's published work *The New and Admirable Arte of Setting of Corne* stressed the currency of its subject, stating that recent years had seen many 'rich experiments' with corn-growing 'farre beyond the hopes and haruests of all our predecessours'.[34] As studies of early modern corn-trading show, supplying London with grain was a vital problem. Proclamations were issued to attract grain from neighbouring counties, millers and bakers were placed under supervision, public granaries were established, city guilds gave money, and still, demand outran supply.[35] This growth in demand also brought home the interrelatedness of the markets for other food and fuel products. London's need for provisions other than corn, which could be satisfied only by a close supplier, compelled it to look further away for its corn supply.[36] Thus, corn productivity was both an influence upon and influenced by metropolitan growth.

Platt responded to these circumstances by outlining measures for the better cultivation of grains and more economical ways of producing commodities such as bread, starch and beer, which competed for the available grain. He consulted his copies of della Porta and Virgil. Adopting della Porta's advice, Platt asserted that the best corn seed was found in the middle ears, and the time-consuming task of picking such seed he allotted to children, so that human resources within a farmer's household were not wasted.[37] He also devised a setting tool – a board three feet long, 12 inches wide, with equidistant wooden dibbers 'of the bignesse of ones finger'. To this time and labour-saving device, he added the caveat that the precise effect of depth and distance upon corn yield must be discovered by trial and error, given the variable factors of climate and soil type.[38]

Describing fertilization by steeping, Platt complained that Virgil, in 'poeticall vaine', recommended the use of salnitre and lees of olive oil as manure in the *Georgics* but did not clarify the pragmatic question of whether the grain was to

---

33 Norman Scott Brien Gras, *The Evolution of the English Corn Market: From the Twelfth Century to the Eighteenth Century* (Cambridge, MA: Harvard University Press 1915; repr. New York, 1967), esp. pp. 9 and 36. Also see Walter, 'The Social Economy of Dearth', pp. 100–103; Andrew Appleby, 'Grain Prices and Subsistence Crises in England and France 1590–1740', *Journal of Economic History*, 39/4 (1979): pp. 865–87; and Edward Anthony Wrigley, 'Some Reflections on Corn Yields and Prices in Pre-industrial Economies', in Schofield and Walter (eds), *Famine, Disease, and the Social Order*, pp. 235–78.

34 Platt, *Arte of Setting Corne*, sig. A3v.

35 Gras, *Corn Market*, Ch. 3; *The Renewing of Certaine Orders Deuised by the Speciall Ccommandement of the Queenes Maiestie, for the Reliefe and Stay of the Rresent Dearth of Graine within the Realme ... 1586. Now to Bee Againe Executed this Present Yere 1594* (London, 1594).

36 Gras, *Corn Market*, pp. 107–9 and 126–7.

37 See della Porta, *Natural Magick*, p. 109; and Platt, *Arte of Setting Corne*, sig. D2v.

38 Ibid., sigs A3v and B1r-B2r.

be steeped or sprinkled with the concoction.[39] Dissatisfied with the imprecision of Virgil's poetry, he moved on to della Porta's discourse, which, despite its 'figuratiue termes', had a practical core:

> … let the bridegrome make choice of such a wife as is of the middle sort, and not of the first or last borne, because they are of a weaker constitution, let them both goe into a warme bath, and there annointing themselues with sweete ointment, and with the fatte food of old Goates, being mixed with Bacchus and Vulcan; let their soft and even layd bed bee gently warmed: for by that viuifying heat they will vnite themselues in an amiable maner, and bee knit and ioyned together in most sweet and louing imbracements: and the seede being thus quickened by the powerfull heate of the man, will in the end bring forth a kindly, and no bastardly issue … But yet I must admonish you of one thing more, that such a wife as is fitte for God Bacchus, must not be bald and without haire, for so wanting the due ornament of her head, shee might happily be despised of her husband, neither should shee haue whereat to purge her owne excrements, onely let all curlings and perriwigs bee laid aside, that so being the more retchless [careless] in her attire, shee may seeme more pleasing and acceptable to her husband.[40]

The mystical marriage, explained Platt, with some annoyance, was 'nothing els but a soking of corne in wine', and he analyzed the particulars of the parable in 'plaine termes': the wife must be of the middle sort because seeds taken from the middle of the ear were best. The bath and ointment meant some 'fat temper' made with goat dung and moistened with wine, producing the heat signified by the term 'Vulcan'. The carefully made bed referred to digging the soil to give it a 'fine and subtill bodie', the bed's warming to the moistening heat of the wine (Bacchus), and the hair to the chaff of the corn which was retained during setting. Saving the reader the trouble of 'beating out the sense of this mysticall Latine', Platt further detailed the procedure of steeping with reference to the required strength of the solution, time and proportion. He also suggested substitutes, including ale, which (rather than wine) was, he explained, the 'naturall bath' of barley and wheat.[41]

In Platt's translation and expansion of measures proposed in *Magia naturalis*, the multiplication of corn became closely linked with the idea of defeating dearth

---

39 Ibid., sig. B2v. Platt quotes Virgil, *Georgics*, Book I, ll.192–99, and expounds the practical import of the passage before asserting: 'But now let us leaue *Virgill* to his poeticall vaine, and let us come to that glorious Neapolitane *Iohan. Baptista Porta* … and washing our eyes first in Eyebright water, let vs see whether we are able to pierce and penetrate into these thicke and foggie clowdes of skill'. Platt maintains this emphasis on clarity, practicality and readers' engagement ('let *us* see') in his analysis of della Porta which follows his exposition of Virgil.

40 Platt quotes from della Porta's *Magiae naturalis* (Frankfurt, 1584) and follows up with his English translation in *Arte of Setting Corne*, sigs B3v–B4r and B4v–C2r.

41 Platt, *Arte of Setting Corne*, sigs C2r–C3r.

by creating, through good husbandry, a near-mythical abundance. If all was done according to the methods outlined, the reader was extravagantly promised 800 corns from one seed. Della Porta, Platt added triumphantly, promised only 200.[42] The exaggeration was in tune with contemporary dreams of plenty at a time of dearth. Such hope was powerfully signalled on the title page of Platt's treatise where the motto 'Adams Toole Revived' accompanied by the illustration of a spade was followed by that of a flourishing wheat plant, below which was written, 'Magnus Deus in minimis'.[43] If the greatness and generosity of God was to be found in small things, corn-growing itself – where abundance flowered from a single seed – became an apt illustration of divine operations. And the methodology of Platt's treatise itself – with its eye on every pragmatic minutia – became an enticing illustration of how attendance to each little mundane detail of practical labour could perhaps realize dreams of plenty.

The Arte of Setting of Corne was by no means the only work by Platt that addressed the contemporary corn crisis. His manuscripts and published works contain numerous instructions for making bread, starch and beer using additives or substitutes for wheat such as peas, beans, beech mast, buckwheat, chestnuts, acorns, vetches, arrowroot, turnips, pumpkins and parsnips. Many of these were initially drawn from both his reading of della Porta's *Magia naturalis* (especially Book 4 on 'Oeconomia'), or the German physician Joachim Strupp's work, *Anchora famis* (1574), and also from his consultations with local acquaintances, practitioners and traders. The advice in books is often compared and blended with the advice obtained from people: for instance, Platt obtained from Bernardino de Mendoza, Spanish Ambassador to France, instructions to make bread from wheat straw. With this, assured Mendoza, he had once relieved a Spanish town 'in an extreame dearth, and scarcity of victual', and showed Platt such a loaf.[44] This practice, Platt later discovered, was usual in some parts of England during hard times. Upon experimenting with the technique himself, he discovered that it made the bread too brown and gritty. The problem, he concluded, lay in English grinding methods – 'our stones be not apt … the same cannot wel be ground but in a steele mill, or hand mil'.[45] By bringing della Porta's secret into an immediate context of pragmatic endeavour and exchange, Platt ensured its revelation: a secret utilized in an actual and current context was a secret demystified.

---

42 Ibid., sig. C3r.

43 Platt is here putting to modified use the familiar Augustinian motto: 'Deus magnus in magnis, maxima in minimis'.

44 Accounts of such crises were not unusual. Giovan Battista Spaccini narrates an extreme case in 1601 in Reggio when three abandoned children of peasants tried to stay alive by consuming boiled straw. See Giovan Battista Spaccini in G. Bertoni, T. Sandonnini, and P.E. Vicini (eds) *Cronaca Modenese (1588–1636)* (Modena, 1911 and 1919), vol. 2, p. 177; and Piero Camporesi, *Bread of Dreams: Food and Fantasy in Early Modern Europe*, trans. David Gentilcore (Oxford: Oxford University Press, 1989).

45 Hugh Platt, *Remedies for Famine* (London: Peter Short, 1596), sig. B2r.

As Platt recorded the results of his trials he occasionally found assistants to help gather and maintain records. In one manuscript, bread experiments are recorded jointly in Platt's own writing and an assistant's hand.[46] Platt takes over from the fourteenth receipt onwards, recording experiments with dearth breads and devising fuel-saving methods of baking. Such manuscript notebooks also exhibit constant concern with alternative staples like macaroni and couscous.[47] These, as we have seen, were specifically marked out as products for 'the poore' and sold from Platt's 'Jewell house'. They also found a place among Platt's lists of mariners' victuals as 'wholesome' and 'nourishing' foods on arduous voyages: as his manuscripts affirm, Platt had taken his earliest versions of these receipts from Francis Drake.[48] In his own modified applications of these receipts, Platt seems keenly concerned about how a single product could find multiple uses and be marketed in different ways. Platt's practical efforts to produce dearth-time bread were painstaking and sustained. Of the 11 receipts in the first part of his published book of remedies for the 1590s 'famine', eight were concerned with bread and the economical use of grains. In the second part of the book, Platt provided an abstract of famine remedies from Strupp's work – of these 40 measures, 29 dealt with making bread cheaply.[49] In such publications, Platt highlighted the economic and ethical purpose of his receipts, emphasizing his aim 'that the bulke and body of our meale and flower will be much increased and multiplied at the least for the poore mannes Table'.[50]

Meticulous attention to increasing and multiplying received knowledge for civic benefit is not only a feature of Platt's extraordinary bread-making and corn-growing activities. A thorough examination of his marginalia, manuscripts and published works reveals that he subjected to this kind of scrutiny a wide range of issues from secrets in medical and alchemical distillations to secrets in grafting and manure. It is far beyond the scope of this chapter to engage in a detailed reconstruction of Platt's

46 Platt, BL, Sloane MS 2189, fols 56–7. The assistant's is an italic hand and it is tempting to speculate that this may have been Platt's wife Judith, who often assisted him with his work.

47 See, for example, Platt, BL, Sloane MS 2216, fol.112v; MS 2189, fols 41, 118, 126 and MS 2244, fol. 29v.

48 Hugh Platt, *Certaine Philosophical Preparations of Foode and Beverage for Seamen* (London, 1607).

49 Platt, *Famine*, Appendix: 'An abstract of certaine frugall notes, or obseruations in a time of Dearth or famine, concerning bread, drink, and meate, with some other circumstances belonging to the same, taken out of a Latin writer, intituling his booke, *Anchora famis & sitis*'. This refers to Joachim Struppe's *EITOPOTIAMATECNIA. Antidotarii antimagistrigi* (Frankfurt, 1574). Struppe (1530–1606) practised medicine in Frankfurt, becoming (in 1575/6) personal doctor, advisor and librarian to Elector Friedrich IV of the Palatinate. The British Library possesses a presentation copy of *Anchora famis* inscribed to Queen Elizabeth I (Classmark: D-7953f.34).

50 Platt, *Famine*, sig. A2v.

process of ideation with respect to all the issues he addressed, but the purpose of this case study of bread and grain is to demonstrate the necessity of undertaking such reconstruction. With his receipts, Platt demonstrated that there were several possibilities for 'turning this our penury into plenty', as he put it, if readers of books of secrets as well as readers of his own books exercised their imagination and simultaneously engaged in hands-on endeavour so that their *active* reading could contribute to the shaping of reliable and economically applicable knowledge.[51] Taken together, the subjects of Platt's investigations suggest his formulation of a programmatic effort to produce such knowledge.

The evolution of this view led Platt to a distinctive form of self-fashioning which tried to abstain from the language and tropes of courtliness and was more suited to Platt's developing ethic of revelation. In the dedication to Anne, Countess of Warwick, in his first publication *Floures of Philosophie* (1572), Platt had used the standard courtier-like trope of the virtuous patron in whose hands Platt's flowers of moral philosophy and poetry would look best, and serve to fan away any external 'noysome smell of vice'.[52] In contrast, the dedication to Lord Essex in *Jewell House* highlights the book's endorsement of applied reading and study suited to the needs of a specific historical moment. Platt argues that 'the trew end of all our priuat labors and studies, ought to bee the beginning of the publick and common good': to this end, he holds himself 'bound by the law of nature, & … the necessity of the times' to 'disclose' the secrets of art and nature, which he had 'long since enterred in a case of marble'. Essex as desired patron is the 'Achillis shield' that protects the Jewell House from 'thundering volees of shot' and breathes life into 'these breathless ghosts of mine, which are thought to bee dead, and rotten longe since within the graues of obliuion'.[53] Platt's hyperbole warns that 'secrets' which are products of long-term labour are easily lost if not disseminated, practiced and refined. Platt's work was flourishing in 1594 when this dedication was published and he was not in immediate danger of rotting in the grave of oblivion personally or professionally; but his posture powerfully announces a movement from secrecy to disclosure.

The brief dedication is followed by a more elaborate preface to the reader where Platt asks:

> Why should not I, hauing spent som of my sweetest hours in reading, & many of them in conference, and more in practise, but most of al in contemplation … adventure as boldlie as the rest, to comend the flowers of my youth, to the courteous view of al well disposed Readers, who may either to their great good make vse of my labors, if they haue been wel bestowed, or else by my example learn to employ both their wittes and time in a course more commendable for themselves, and more profitable for their Countrey?[54]

51 Ibid., sig. A2r.

52 Platt, *Floures of Philosophie*, Dedication, sig. A2v.

53 Platt, *Jewell House*, sig. A2r-v.

54 Platt, *Jewell House*, sig. B4v.

The very structure of the rhetorical question replicates the process of disclosure: the reader of the preface walks with Platt through the progressive enrichment of his 'sweetest hours' of reading by conference, practice and contemplation. But by the end of this cunningly constructed sentence, Platt has deftly released his reader's hand and left him/her with the sense of being offered a legacy that could only be preserved by following Platt's own example of active reading.

Arguably, the most powerful analogy for such reading available to Platt was derived from contemporary theatre, and he concludes his preface to *Jewell House* in the familiar manner of a Renaissance playwright concluding the prologue to his play:

> But now it is high time for the prologue to giue place because the Actors are at hand, who are readie to present such choice and varietie of matter, as that notwithstanding they may happily faile in gesture or action, yet I doubt not but that they wil either procure a friendly & thankefull plaudite, which is the most that I can desire, or a most free & liberal pardon, which is the least that I can deserue.[55]

Theatricality was an important element in the self-fashioning of the professors of secrets on whom Platt so often draws.[56] But Platt's particular sleight of hand is one that uses theatrical tropes to secure the link between reading, knowing and doing which is so crucial to his critique of secrecy. He affects, as the sentence quoted above shows, a kind of self-effacement by signalling his own timely withdrawal from the stage. The ending of the prologue is in fact an opening for the 'jewells' (or receipts) which are enlivened as actors and agents performing gestures and actions authored by Platt. This enables a shrewdly literalized and physicalized evocation of the all-too-familiar stage motto, 'Totus mundus agit histrionem'. Platt's animated jewels then draw the audience (or readers) onto the 'stage' by demanding their mental and physical engagement with methods and causes and by compelling them to be labourious as well as ingenious and inventive participants or performers. Platt's stance subtly renovates the conventional notion of the stage as a vehicle for wonder and artifice: he draws attention to its being, in fact, a locus of physical cognition which made knowledge both transferable and active, and thereby dismantled 'wonder'. Moreover, by Platt's astute placement of this rhetorical device at the end of the preface/prologue, the readers/audience are able to view the receipts which immediately follow on the adjacent page as sites of active engagement and understanding. 'Reading is performing; knowing is doing', Platt's personified 'jewels' seem to say as they tempt readers, in 'plain and familiar terms', to transform words into things.[57] Such engagement and 'experience' was,

---

[55] Ibid.

[56] See Eamon, *Science*, p. 227.

[57] For a discussion of the seventeenth-century theatre as 'the perfect vehicle for an age obsessed with the relationship between words and things', see Pamela H. Smith, *The*

as Platt wrote, 'the undoubted mother of all true and *certaine* knowledge … kept in the bosome of Nature'.[58]

In his later years, with financial troubles and ill-health looming large, Platt grew increasingly anxious about his ability to continue to disclose or 'disperse' the 'most secret Iewels' in 'Natures Cabinet'. He complained that without active assistance from the authorities, his critique of secrecy would go unnoticed, the knowledge he had unlocked would not be disseminated and Nature's cabinet would remain closed to the public.[59] But this fear of closure was not entirely confirmed – or so it seems, from the frequent republication of Platt's works over the next century,[60] from the repetition of Platt's ideas by Francis Bacon as the latter more famously formulated his 'active science',[61] from the adoption of Platt's methods and measures by the Hartlib circle,[62] and indeed from the marginalia produced by early modern readers in surviving copies of Platt's own published books.[63] The first modern edition of Platt appeared during the post-war crisis in 1948 with the editorial claim that this early receipt book (*Delights for Ladies*, 1609) would disclose the lost knowledge of principles of domestic economy and health that was especially needed in the shortage-driven

---

Business of Alchemy: Science and Culture in the Holy Roman Empire (Princeton, NJ: Princeton University Press, 1994), pp. 266–71. It is also worth noting that the fraught relationship between words and things is a frequently debated theme among early modern English playwrights themselves: especially powerful examples may be found in the plays of John Ford (1586–c. 1639).

58 Platt, *Diuerse New Sorts of Soyle* (published in *Jewell House*, 1594, with separate title page), sig. A2r.

59 See, for instance, *Certaine Philosophical Preparations of Foode and Beverage for Sea-men*, published in 1607, the year before Platt's death.

60 Platt's *Jewell House*, first published in 1594, was reprinted in 1653 with an appendix by Arnold de Boot; while *Floraes Paradise* was reprinted seven times with the title *The Garden of Eden* in 1652, 1653, 1654, 1655, 1659, 1660 and 1675. These last two editions contained a 'second part' reproduced from Platt's manuscripts.

61 See Webster, *The Great Instauration*, p. 468; Eamon, *Science*, pp. 311–14 and 319–50. For a recent comparison of Platt and Bacon, see 'From the Jewel House to Salomon's House' in Harkness, *Jewel House*, pp. 211–53.

62 Samuel Hartlib's 'Philosophical Letter Concerning Vegetation or Causes of Fruitfulness', emphasizes the work of 'learned Philosophers', such as Hugh Platt, in *His Legacy of Husbandry* (London, 1655), pp. 217–19; Daniel Coxe, 'Enquiries Concerning Agriculture', *Philosophical Transactions of the Royal Society* 1 (1665): pp. 91–4; Review of Platt's *The Garden of Eden*, *Philosophical Transactions of the Royal Society* 9 (1675): pp. 302–4.

63 Early modern marginalia appear in many extant copies of Platt's published works. An appropriate example in the context of this essay would be a copy of *Arte of Setting of Corne* (1600) held in the British Library (STC, 2nd ed./19993; BL classmark: C.54.b.28) which contains elaborate calculations of possible yields from corn (based on chapter 5 of the text) and other crops. See esp. sigs A1v, B2r, D3r, and D4r onwards.

households of post-war England.[64] So Platt's 'secret jewels', it seems, have been dispersed at various opportune moments; it is Platt, as author and critic, who remains relatively little-known and still hovers in the wings.

[64] See editorial preface by George E. Fussell and Kathleen Rosemary Fussell in Platt's *Delights for Ladies* (1609), facsim., published in post-war London in 1948.

Chapter 4
Robert Boyle and Secrecy

Michael Hunter*

Our image of Robert Boyle has changed markedly in recent years. This is especially true in relation to alchemy, since, thanks particularly to the work of Lawrence M. Principe, we are aware that Boyle was more fully involved in alchemical pursuits than previous historians had either realized or been happy to accept.[1] We also now have a better understanding of Boyle's complex personality, in which anxious soul-searching coexisted with acute concern about the esteem in which he was held by others and about his relations with the wider world.[2] But, though Boyle was to some extent a victim of traits of the latter kind, he was also a manipulator, who could use these and other facets of his personality and status to his advantage, which further complicates matters. The result is that, in the twenty-first century, Boyle has become a more mixed-up and perhaps therefore more interesting figure than the rather lifeless lay saint depicted in the traditional historiography.

On the other hand, though such hitherto neglected facets of Boyle have complicated the traditional view of him, many elements of that view have endured scrutiny. In particular, we need to do justice to the programme in natural philosophy which he so influentially expounded in such writings as *Certain Physiological Essays* (1661), including his insistence on the need for a full account of experimental and other findings as a basis for improved understanding of the natural world, and his championing of the mechanical philosophy. All this formed part of a project for the progress of science both in its own right and for its potential religious and utilitarian spin-offs which Boyle powerfully promoted in his profuse writings over many years. The challenge is to give an account of Boyle which does justice both to the aspects of him that have long been familiar and to the complications introduced by recent scholarship.

---

* I am grateful to Peter Anstey and Lawrence M. Principe for their comments on a draft of this chapter, to Victor Boantza for showing me the paper referred to in note 53 prior to publication and to Will Poole for the further reference included in the same note.

1 See especially Lawrence M. Principe, *The Aspiring Adept: Robert Boyle and his Alchemical Quest* (Princeton, NJ: Princeton University Press, 1998).

2 See Michael Hunter, *Robert Boyle (1627–91): Scrupulosity and Science* (Woodbridge: Boydell Press, 2000), and Hunter, *Boyle: Between God and Science* (New Haven, CT and London: Yale University Press, 2009).

Boyle's attitude to secrecy is a key test case in this respect, since the relationship to the secrets tradition of as pivotal figure in the science of his generation as Boyle is obviously worth understanding in its own right. Boyle is all the more interesting because of the extent to which he spans the various contexts in which secrecy typically arose at the time – from the fully-fledged alchemical tradition in which he participated, with its elaborate protocols of secrecy, encryption and mystification, to the craft tradition and its subset in the form of medical practice, where there was an incentive to keep trade secrets because practitioners' livelihoods depended on them. In Boyle's case, a related issue on which he had strong views concerned authorship and intellectual property, since he believed that the originator of a discovery deserved proper credit for it. There is also the context of virtuoso values, in which secrecy overlapped with the analogous concept of 'rarity', phenomena being valued for their wonderful and inexplicable qualities, and their ownership seen as a mark of exclusivity.[3]

What is today probably Boyle's best known writing on secrets is one which has in fact only been generally known since 1950, when it was reprinted by Margaret Rowbottom: since then, however, this has often been seen to epitomize the call for openness which was the hallmark of the Baconian scientific tradition of his day.[4] The work in question is Boyle's 'Invitation to a free and generous Communication of Secrets and Receits in Physick', published in *Chymical, Medicinal, and Chyrurgical Addresses: Made to Samuel Hartlib, Esquire* in 1655. In it, Boyle made an impassioned plea for the free dissemination of knowledge, especially medical, answering a range of objections put forward by those who advocated the withholding of useful recipes on the grounds of the prestige that accrued to things for their rarity, or the danger that they might be spoiled in the course of being made common. Boyle's response was that those in possession of secrets that might be of general benefit were obliged on both religious and moral grounds freely to divulge them. He took the view that, even when people had accepted a recipe on condition of secrecy, their duty was to violate this.[5]

Ironically, we now know that this work was compiled at a date which precedes Boyle's hands-on experience of natural philosophy. In his early adult years in the late 1640s Boyle was surprisingly little interested in science, instead

---

[3] For a helpful recent overview of virtuoso values, see Brian Cowan, *The Social Life of Coffee: the Emergence of the British Coffeehouse* (New Haven, CT and London: Yale University Press, 2005), esp. pp. 10ff. On Boyle and intellectual property see Hunter, *Scrupulosity and Science*, esp. pp. 219–21.

[4] See, for example, William Eamon, 'From Secrets of Nature to Public Knowledge', in David C. Lindberg and Robert S. Westman (eds), *Reappraisals of the Scientific Revolution* (Cambridge: Cambridge University Press, 1990), pp. 333–65 (p. 351), and Jan Golinski, 'Chemistry in the Scientific Revolution: Problems of Language and Communication', in ibid., pp. 367–96 (p. 383).

[5] *The Works of Robert Boyle*, ed. Michael Hunter and Edward B. Davis (14 vols, London: Pickering & Chatto, 1999–2000), vol. 1, pp. 1–12.

writing treatises on ethical theory and related texts which he hoped would help to reform the morals of his benighted peers. The 'Invitation' forms part of this group of writings by Boyle and it fits into this tradition in its essentially moralistic stance: though it clearly reflects Boyle's experience as a consumer of medications, it is arguably rather limited in its sympathy for the views of their producers.

Secrecy in Boyle's Early Natural Philosophy

When we look at Boyle's attitudes in the period *after* he discovered natural philosophy around 1650, we find him displaying typical virtuoso attitudes, including, perhaps not surprisingly, the classic language of secrecy. This is seen in the workdiaries that he compiled in these years, which provide a rather breathless account of his pursuit of medical recipes and other material from fellow enthusiasts, and which often evaluate the information that he obtained in such terms. Thus Boyle explained how his newly-made acquaintance, Sir Kenelm Digby, 'commends above all his Secrets' a formula which he had obtained from Basil Valentine's *Currus triumphalis*, while of another of Digby's recipes Boyle noted: 'This he tells me is his Greatest secret in perfuming'.[6]

Much the same is true of Boyle's letters in these years, for instance one to his Oxford acquaintance, Ralph Bathurst, in 1656, in which, acknowledging receipt of a recipe for distilling spirit of roses, Boyle wrote: 'The secresy that you command in reference to the receipt, its owne nature would have exacted from me'.[7] The notes about Boyle in Samuel Hartlib's 'Ephemerides' tell a similar story. Thus Hartlib recorded how Boyle accounted a recipe for essence of jasmine 'one of the greatest secrets', while, in telling of a friend of Boyle's who had a method of making apples grow unusually large, he added: 'but Mr Boyle is bound to secrecy and binds others to it'.[8]

By this time, Boyle had been exposed to an apprenticeship both in high-level chemical experimentation and in the secrets tradition by one of the classic exponents of both, George Starkey, the American alchemist who came to England in 1650 and almost immediately struck up a friendship with Boyle. Starkey's profuse letters to Boyle in the year after they met are full of news of secret processes which he had discovered which he vouchsafed to Boyle only on condition that he kept them to himself. In fact, we now know that Boyle was spectacularly bad at doing so, that (in the words of William R. Newman and Lawrence M. Principe) he

[6] The Workdiaries of Robert Boyle. Available at www.livesandletters.ac.uk/wd/index.html (hereinafter WD) 12-5 and 13-6 (citations refer to workdiary and entry numbers).

[7] *The Correspondence of Robert Boyle*, ed. Michael Hunter, Antonio Clericuzio and Lawrence M. Principe (6 vols, London: Pickering & Chatto, 2001), vol. 1, p. 204.

[8] Hartlib Papers, University of Sheffield, 29/7/16A, 29/8/11B.

'appears to have been a constant "leak" of privileged information'.[9] Thus in 1653 we find him vouchsafing one of Starkey's choicest secrets – for the preparation of wine from fermented vegetables – to another of his chymical friends, Frederick Clodius, Hartlib's son-in-law, with the rider: 'Having received this Processe from a freind as a Secret; I shall beg you would not let it loose that Name'.[10] This formula does not suggest that Boyle was reverting to the ethos of openness that he had advocated in his 'Invitation'. Rather his behaviour seems to combine a formal acceptance of the protocols involved with a certain casualness about observing them, possibly as if he considered that his superior social status placed him above such constraints.

A comparable ambivalence is in evidence in Boyle's practice concerning the recipe known as *ens veneris*, a copper compound with powerful medicinal qualities which he and Starkey were inspired to prepare on the basis of a joint reading of J.B. van Helmont.[11] Increasingly, Boyle took credit for this nostrum, alluding only casually to Starkey's role in its discovery, but his attitude towards its dissemination fluctuated. His protégé, Henry Oldenburg, obtained 'the substance of the processe' from Boyle in 1659 while on his continental travels with Boyle's nephew, Richard Jones, and he almost immediately offered it to the Dresden chemist, Gansland. In doing so, however, he specifically stated that the originator of the nostrum, Boyle, had made it a condition 'that I communicate it to no one, but such as have first faithfully promised not to reveal this secret, reserving it for their own use and practice only'.[12] Boyle himself, on the other hand, clearly purveyed the medication widely to those who needed it, while in 1663 he published full details of it in his *Usefulness of Natural Philosophy*.[13]

A further curious episode of about the same period involves an exchange of letters between Boyle and that classic exponent of genteel virtuoso values, John Evelyn. Evelyn had sent Boyle a recipe for varnish, but he explained his ambivalence over publishing such recipes lest he 'debase much of their esteeme, by prostituting them to the Vulgar'. In his reply, Boyle expressed the hope 'that some Expedient may be found to reconcile the disclosure of many secrets, with the keeping up & secureing the Reputation of Learning'. His wording is slightly puzzling. In part, it echoed Evelyn's, who had outlined a plan for an academy at

---

9 William R. Newman and Lawrence M. Principe, *Alchemy Tried in the Fire* (Chicago, IL: University of Chicago Press, 2002), pp. 265–7 (p. 265).

10 Boyle, *Correspondence*, vol. 1, p. 149.

11 See William R. Newman, *Gehennical Fire* (Cambridge, MA: Harvard University Press, 1994), pp. 71–2, and Newman and Principe, *Alchemy Tried in the Fire*, pp. 9 and 221–2.

12 Boyle, *Correspondence*, vol. 1, pp. 328–9; *The Correspondence of Henry Oldenburg*, ed. A. Rupert Hall and Marie Boas Hall (13 vols, Madison, WI and London: University of Wisconsin Press, 1965–86), vol. 1, pp. 243–4. I owe the latter reference to David Haycock.

13 Hunter and Davis, *Works*, vol. 3, pp. 391–2, 500–5.

which well-born youth would be schooled in virtuoso values '(not without an Oath of seacresy)', and the superintendent of which would from time to time divulge specimens of such material 'for the reputation of Learning, and benefit of the Nation'. But Boyle's take seems perceptibly different, reflecting his developing agenda for demarcating between 'experimentall & notionall Learning', and his view that the best way to encourage his fellow countrymen to value the former was to disseminate 'reall & usefull Productions' which exemplified its potential.[14]

By this time, we have reached the period in the late 1650s when Boyle developed the distinctive outlook in natural philosophy for which he has since been remembered, writing a notable series of natural philosophical treatises which were published in the 1660s and established him as one of the most influential exponents of the 'new' empirical science of his period. In these, as in his comment to Evelyn, we find echoes of the ethos of his early 'Invitation to Communication', and it important to remember that Boyle never lost the moral imperative of his early writings, simply transferring it to new fields, particularly the study of nature. On the other hand, it was supplemented by new criteria which reflected the natural philosophical agenda which he was expounding at this time and which had been missing before.

Boyle's Statements on Secrecy

What, therefore, did Boyle say about secrecy in the writings that he compiled at this point, in which he laid down clear protocols for the pursuit of science, in terms both of its explanatory goals and of the most appropriate methods to achieve them? Boyle included various statements on the issue, which we must consider before turning to the evidence of Boyle's actual practice in relation to secrets in these and other writings.

First, there is *The Sceptical Chymist* (1661) and Boyle's familiar strictures on the mode of writing associated with the chemists whom he attacked in that work. He wrote how:

> though much may be said to Excuse the Chymists when they write Darkly, and Ænigmatically, about the Preparation of their *Elixir*, and Some few other grand *Arcana*, the divulging of which they may upon Grounds Plausible enough esteem unfit; yet when they pretend to teach the General Principles of Natural Philosophers, this Equivocall Way of Writing is not to be endur'd. For in such Speculative Enquiries, where the naked Knowledge of the Truth is the thing Principally aim'd at, what does he teach me worth thanks that does not, if he can, make his Notion intelligible to me, but by Mystical Termes, and Ambiguous Phrases darkens what he should clear up; and makes me add the Trouble of guessing at the sence of what he Equivocally expresses, to that of examining the

[14] Boyle, *Correspondence*, vol. 1, pp. 212–5.

> Truth of what he seems to deliver. And if the matter of the Philosophers Stone, and the manner of preparing it, be such Mysteries as they would have the World believe them, they may Write Intelligibly and Clearly of the Principles of mixt Bodies in General, without Discovering what they call the Great Work.[15]

As will be seen, despite Boyle's acknowledgment that there were certain secrets which it was perfectly legitimate for chemical writers not to divulge, he was equally emphatic that openness and clarity were appropriate in relation to natural philosophy, and elsewhere he was to express disdain for those who possessed knowledge but '(Chymist like) keep it secret'.[16]

A second relevant statement appears in the 'Proemial Essay' to *Certain Physiological Essays* (1661) where Boyle addressed the issue of 'why in the ensuing Essays I have mention'd divers Experiments which I have not plainly and circumstantially enough delivered'.[17] In this case Boyle's response relates to secrecy in the context of craft practices: after first apologizing that he may have been unduly succinct in expounding some processes, he went on to explain how he had 'purposely omitted some manual Circumstances', justifying this partly out of solicitude for tradesmen whose livelihood depended on such products, and partly by demarcating between what was appropriate in natural philosophy and what was appropriate in other fields in a manner which echoes *The Sceptical Chymist*. Even more pertinent are Boyle's further considerations:

> Thirdly, I mention'd some things but darkly, either because I receiv'd them upon Condition of secrecy, or because some ingenious persons that communicated them to me, or others to whom I imparted them, do yet make, and need to make, a pecuniary advantage of them. Fourthly, And some things that, either having been the fruits of my own Labours, or obtain'd in Exchange of such, are freely at my own disposal, I have not yet thought fit so plainly to reveal, not out of an envious design of having them bury'd with me, but that I may be always provided with some Rarity to barter with those Secretists that will not part with one Secret but in Exchange for another, and think nothing worth their desiring that is known already to above one or two Persons. And I think it very lawful to reserve always some conceal'd Experiments by me, wherewith to obtain the secrets of others, which being thereby gain'd, the other (as being no longer necessary to the former end) may freely be communicated.

It is interesting how the last phrase actually echoes the earlier 'Free Invitation', as does his disavowal concerning secrets of 'an envious design of having them bury'd with me'. Equally notable is the reference to his practice of 'bartering' recipes: this bears out the note of the physician and memorialist, John Ward, on

15 Hunter and Davis, *Works*, vol. 2, p. 292.

16 Ibid., vol. 5, p. 418.

17 Ibid., vol. 2, pp. 30–31.

a visit to Boyle in 1667 that he 'hardly parts with any mony to buy Receipts, but only exchanges', and a similar line was later taken by the physician and naturalist, Martin Lister, who told Oldenburg that he considered such exchange 'the most creditable & proper use of secretts'.[18] This looks forward to Boyle's actual practice in ways that I will come to shortly, the implication being that Boyle 'concealed' experiments only in order to swap them for others and thus make it possible freely to communicate both – though it might be felt that this itself undercut the ethos of secrecy, since the effect of such practice would be to devalue the very item that Boyle had just bartered.

A further reflection on related issues occurs in Boyle's 'Preamble' to Part 2, section 2 of *The Usefulness of Natural Philosophy*, mainly written in the late 1650s although not published with the 'First Tome' of that work in 1663, instead seeing the light of day only in 1671.[19] This passage is interesting not least because Boyle here shows his awareness of the books of secrets tradition, actually alluding to authors like 'Cardan, Weckar, and Baptista Porta'. In it Boyle sought to answer the 'objection, whether I doe not injure Tradesmen by discovering so plainly those things, which our Laws call the Mysteries of their Arts', claiming firstly that 'I never divulge *all* the Secrets and practices necessary to the exercise of any one Trade, contenting my self to deliver here and there upon occasion some few particular Experiments, that make for my present purpose'. Secondly, he argued that 'it is not the Custome of Tradesmen to buy Books', whereas those who typically purchased these were ill-equipped to implement the technical prescriptions that they contained, not least since this usually required 'a Manuall dexterity' which was 'not to be learnt from Bookes, but to be obtaind by imitation and use'. But he went on to claim that:

> though some little inconvenience may happen to some Tradesmen by the disclosing some of their Experiments to practicall Naturalists, yet that may be more than compensated, partly, by what may be contributed to the perfecting of such experiments themselves, and, partly by the diffused Knowledge and sagacity of Philosophers, and by those new Inventions, which may probably be expected from such persons, especially if they be furnished with Variety of hints from the practices already in use. For these Inventions of ingenious heads doe, when once grown into request, set many Mechanical hands a worke, and supply Tradesmen with new meanes of getting a livelyhood or even inriching themselves.

Boyle then proceeded to give examples of the way in which tradesmen might benefit from the superior inventions which naturalists produced, instancing the

---

18 Folger Shakespeare Library, Washington, DC, MS V.a.296, fol. 20; Hall and Hall, *Oldenburg*, vol. 10, p. 383.

19 Hunter and Davis, *Works*, vol. 6, pp. 397ff. For parallel passages, see ibid., vol. 4, pp. 114, 175 and 180.

profitable trade in optical instruments which London instrument-makers like Richard Reeves enjoyed on the basis of scientific improvements going back to Galileo.

Again, this is a revealing passage, notable for its condescension towards humble operatives in comparison with the lofty status accorded to natural philosophers on the basis of the analytical understanding to which they aspired. What was the source of this? In part, it stemmed from the strong faith in the potential of experimental philosophy to provide a true understanding of nature which formed so influential a part of Boyle's philosophical outlook. But where did *this* come from? Here we may perhaps discern the extent to which Boyle transposed to his role as a natural philosopher the attitudes associated with his privileged status as one of the richest aristocrats in England, including the instinctive expectation of deference from his social inferiors. Boyle displayed similar attitudes in his encounter with the world of projecting when he became involved in a scheme for desalinizing sea-water in the 1680s and when he seems similarly to have presumed that his ability to understand the rationale of processes gave him a right simply to ignore the interests of rivals.[20]

A final general reflection by Boyle on the pros and cons of divulging secrets had appeared in the first section of Part 2 of *The Usefulness of Natural Philosophy*, published in 1663, where, prior to giving the recipe for the medicine, 'Pilulæ Lunares', he apologized for not naming the 'very Ancient and experience'd Chymist' from whom he had derived it, on the grounds that it was often 'more prejudicial then grateful to one that makes an advantage by the Practise of Physick, to annex in his lifetime his name to some of his Receipts or Processes', since this encouraged people to circumvent him in various ways.[21] This opened up a whole can of worms about divulging 'communicated recipes' in medicine, Boyle's convolutions concerning which I have expounded elsewhere.[22] Overall, however, it will be seen that, through such programmatic statements, Boyle was provided with a range of justifications on the one hand for divulging secrets, but on the other for being somewhat circumspect in doing so. Indeed, it could be argued that he was thus provided with the potential for a considerable amount of flexibility in practice, enabling him to justify a range of strategies as it suited him. What, therefore, did he actually do?

Boyle's Attitude in Practice

It is certainly the case that, in Boyle's time as earlier, there was a clear expectation fostered by the secret tradition that it would have been legitimate for Boyle to

---

20 See Hunter, *God and Science*, p. 216. For background on the role of status see Steven Shapin, *A Social History of Truth* (Chicago, IL: University of Chicago Press, 1994).

21 Hunter and Davis, *Works*, vol. 3, pp. 485–6.

22 Hunter, *Scrupulosity and Science*, Ch. 9.

keep things secret had he so wished. This is seen in those of his correspondents who presumed that he might be reluctant to divulge details of recipes, asking for information only 'if not kept as a secret', or promising that 'you may engage me to secrecy in what you please'; Henry Oldenburg similarly assured Boyle concerning data that he received from him that 'you may be sure, I shall keep them secret, as long as I receive no leave to divulge them'.[23] Boyle was certainly capable of withholding information when he wanted to, as part of a repertoire of authorial apologies of which his works are full and which illustrate a perceptible degree of self-indulgence on his part.[24] Thus he even occasionally offered to supply missing information by word of mouth, while he often excuses himself for failing to publish matters in full because he was not yet satisfied with his investigations or on other, comparable grounds, in one case explaining more enigmatically how 'cogent considerations forbid me at present to publish' a process.[25]

Boyle also commonly invoked the ethos of secrecy on the grounds that he had been given details of a practice on condition that he did not divulge its rationale. Thus in connection with a method of staining marble red he explained how this had been revealed to him 'upon condition of secresy (which I have to this day inviolably kept)'. Elsewhere he tells of 'a strange way of preserving Fruits, whereby even *Goos-berries* have been kept for many Moneths, without the addition of Sugar, Salt, or other tangible Bodies', owned by 'an eminent Naturalist, a Friend of yours and mine', adding: 'but all that I dare yet tell you, is, That he assures me his Secret consists in a new and artificial way of keeping them from the Air'.[26] In at least some such cases, it is quite possible that this ostensible solicitousness for Boyle's informants was in fact a subterfuge for protecting secrets of Boyle's own.

In terms of Boyle's actual practice in relation to secrets, on the other hand, the very cases cited above are revealing. In the first of them, Boyle explained how he helped the 'ingenious Stone-cutter in *Oxford*' whose secret it was how 'to improve his Invention by making it practicable with other Colours', thus illustrating an element of interchange that seems to have characterized many such transactions on Boyle's part. The second, on the other hand, has an element of provisionality about it, implying that the status of the secret was quite contingent and that Boyle's ambition was to demystify it.

Indeed, in many ways the latter instance does seem to provide the key to Boyle's attitude on such matters. Since he saw the goal of natural philosophy as to furnish explanations, the ultimate challenge was to make all secrets unmysterious. The same is equally true of phenomena that he described as 'rarities', where Boyle is to be found taking a similar attitude, in this case sometimes destroying items treasured by connoisseurs in order to elucidate their

---

23 Boyle, *Correspondence*, vol. 3, pp. 259 and 344; vol. 4, p. 121.

24 See Hunter, *Scrupulosity and Science*, Ch. 7.

25 Hunter and Davis, *Works*, vol. 3, p. 510; vol. 7, p. 383. See also ibid., p. 387; and, for example, vol. 2, p. 373; vol. 3, p. 412; vol. 5, p. 407 and vol. 10, p. 382.

26 Ibid., vol. 3, p. 358 and vol. 10, p. 139.

physical properties.[27] In the case of secrets, Boyle took it for granted that they all could ultimately be 'cracked', either by obtaining information from the person whose secret it was, or by investigations of his own, and on various occasions he gives interesting hints as to how he went about this. Sometimes, as in the case just cited, he gives a kind of provisional account of a secret, implying that in time he would be in a position to elucidate it, while elsewhere he gives examples of secrets which he obtained by exchange and then de-mystified. Thus he tells of a secret that he obtained from

> a not unlearned Emperick, who was exceedingly cry'd up for the Cures he did, especially in difficult Distempers of the Brain, by a certain Remedy, which he call'd sometimes his *Aurum Potabile*, and sometimes his *Panacæa*; and having obtain'd from this Man, in exchange of a Chymical Secret of mine he was greedy of, the way of making this so celebrated Medicine, I found that the main thing in it was the Spirit of Soot, drawn after a somewhat unusual, but not excellent manner; in which Spirit, Flowers of *Sulphur* were, by a certain way, brought to be dissolv'd, and swim in little drops that look'd of a golden colour.[28]

On other occasions, it seems likely that Boyle conducted his investigations by a mixture of inquiry and experiment. Concerning one such 'choice Secret', he explained how, 'though I could not learn a considerable Particular or two, which belong to the Delicacy of it; yet (partly by putting Questions, and partly by some Tryals of my own) I attain'd to the substance of this Mystery, as they call it, which seems to be this': he then went on to outline how by the application of a coating of wax to an exemplar written in a special ink, fine lettering could be produced on an engraved plate.[29] Perhaps the most interesting example of this mixture of interrogation and experimentation concerns the noctiluca, the phosphorescent substances brought to the attention of Boyle and other virtuosi of the Royal Society by J.B. Krafft and others from 1677 onwards and which Boyle wrote up in his *The Aerial Noctiluca* (1680). This was a typical virtuoso display which capitalized on its strangeness and wonder, while Krafft and the other originators of the substance were careful to protect their rights by keeping the method of preparing it secret.[30] Boyle, however, tried his usual methods. He explained how:

---

27 For example, ibid., vol. 7, pp. 25–6.

28 Ibid. vol. 3, p. 384. For other references to the use of exchange, see, for example, ibid., p. 385 and vol. 13, p. 248.

29 Ibid., vol. 6, pp. 499–500.

30 For an account of the whole episode, see Jan Golinski, 'A Noble Spectacle: Phosphorus and the Public Cultures of Science in the Early Royal Society', *Isis*, 80 (1989): pp. 11–39. See also J. R. Partington, 'The Early History of Phosphorus', *Science Progress*, 30 (1936): pp. 402–12.

> having at Mr. *Kraffts*'s desire, imparted to him somewhat that I discover'd about uncommon *Mercuries*, (which I had then communicated but to one Person in the World) he, in requital, confest to me at parting, that at least the principal matter of his *Phosphorus*'s, was somewhat that belong'd to the Body of Man. This intimation, though but very general, was therefore very welcome to me, because, though I have often thought it probable, that a *shining substance* may, by Spagyrical Art, be obtain'd from more kinds of Bodies than one: yet designing, in the first place, to try if I could hit upon such a *Phosphorus* as I saw was preparable, the Advertisement sav'd me (for some time) the labor of ranging among various Bodies, and directed me to exercise my industry in a narrower compass.[31]

Boyle thereupon pursued this line of inquiry, leading him to focus on urine as the bodily substance from which phosphorus could be made, but he reached an impasse because he was unable to work out the temperature at which the conglomerate needed to be heated to separate out the impurities that it contained. He went on:

> However, adhering to the first choice I had made of a *fit* matter, I did not desist to work upon it by the ways I judg'd the most hopeful, when a learned and ingenious Stranger, (*A. G. M. D.* Countreyman, if I mistake not, to Mr. *Krafft*) who had newly made an Excursion into *England*, to see the Countrey, having, in a Visit he was pleas'd to make me, occasionally discoursed, among other things, about the *German Noctiluca*, whereof he soon perceiv'd I knew the true matter, and had wrought much upon it. He said something about the degree of Fire, that made me afterwards think, when I reflected on it, that *that* was the only thing I wanted to succeed in my endeavors.[32]

In other words, Boyle found that, by combining hints that he was given by their proprietors with his own expertise, he could almost invariably get to the bottom of any secret that presented itself.

Hence perhaps we are coming back by a different route to the ethos of Boyle's early 'Invitation to Free Communication'. Boyle seems to have it taken for granted that secrecy was a contingent status which it was appropriate to try to break down through investigation by one means or other. In a number of cases, his sense also becomes apparent that the demands of natural philosophy and humanity, particularly for secrets which might be of great public benefit, overruled the selfish interests of those who might otherwise have taken their nostrums to the grave with them. On one occasion, he himself acted as the intermediary in this, when he was able to divulge a method of increasing the yield of corn which he obtained from its originator, 'who was pleas'd to make choice of me to instruct his Secret with, that

[31] Hunter and Davis, *Works*, vol. 9, p. 273.

[32] Ibid., pp. 273–4. This was probably Ambrose Godfrey Hanckwitz, later Boyle's operator.

in case he dyed before me, the publick might not loose it'.[33] On others he failed, telling wistfully of excellent nostrums that were lost to posterity because of the selfishness of their proprietor, as in an early workdiary record from Boyle's Irish colleague, Gerard Boate, who informed him of a 'Bohemian Physitian, (whose name I cannot readily remember,)' who had an infallible cure for all sorts of agues which worked within 24 hours: 'he added that this Secret dy'd with the Inventor'.[34]

What can only be described as a similar moralism is in evidence in certain cases where Boyle took the view that inventions or phenomena were best *not* divulged, since these were often powerful processes which he believed could have a harmful effect if placed in the wrong hands, and which therefore 'my love of Mankind has oblig'd me to conceal, even from my nearest Friends'.[35] Indeed, this sense of a threat to the good of mankind is virtually the only grounds on which Boyle states that he was permanently withholding information. He took a similar line concerning a technique which could be used for counterfeiting coins: 'the fear of teaching bad men a skill that probably they will not otherwise acquire, makes me forbear to mention it'.[36]

A similar consideration potentially arose in relation to alchemy. Thus, in a manuscript set of headings for and against divulging the arcanum to which I first drew attention and which William Newman has since published in full, Boyle cited as an argument against such revelation: 'That it would much disorder the affairs of Mankind, Favour Tyranny, and bring a general Confusion, turning the World topsy turvy'.[37] This was echoed by Isaac Newton in the letter that he was stimulated to write to Henry Oldenburg by the article that Boyle published in *Philosophical Transactions* in 1676 about the incalescence of mercury, in which he criticized the author for publicly divulging matters which 'may possibly be an inlet to something more noble, not to be communicated without immense dammage to the world if there should be any verity in the Hermetick writers'.[38]

Secrecy and Alchemy

Turning, therefore, to Boyle's concern with alchemy, this was clearly something of a special case – hardly surprisingly in view of the fact that, even in the passage

33 Ibid., vol. 6, p. 417.

34 WD 6-14. Cf. Hunter and Davis, *Works*, vol. 3, pp. 337, 449 and 538.

35 Ibid., vol. 9, p. 281.

36 Ibid., vol. 8, pp. 538–9. For further comparable references, see ibid., vol. 6, p. 430; vol. 9, p. 293 and Royal Society of London (hereinafter RS) MS 187, fol. 166.

37 Hunter, *Scrupulosity and Science*, p. 114 and Newman, *Gehennical Fire*, pp. 254–5. Since such arguments are balanced by contrary ones, the overall import of the document is ambiguous.

38 *The Correspondence of Isaac Newton*, ed. H.W. Turnbull et al., (7 vols, Cambridge: Cambridge University Press, 1959–77), vol. 2, p. 2.

in *The Sceptical Chymist* already cited in which Boyle criticized the obscurity of chymists' language, he agreed that it was legitimate for the higher arcana to be withheld. In fact, this is an area where Boyle *was* circumspect. A good example is provided by a letter that he wrote in March 1680 to the New England physician, William Avery, who had been inspired by Boyle's reputation to make contact with him. Initially, they discussed medical matters, including the efficacy of herbs found in the New World but not the Old, but when Avery plucked up courage to ask Boyle for information about the Helmontian alkahest, he received the following reply:

> I perceive you have had a very partial representation of my skill in Chymistry & Physick, to the latter of which I do not pretend, nor to the former any further than as an innocent Recreation & an Instrument to discover the nature of things. Therefore you much misaddress yourselfe when you would learn of me the preparation of such high Arcana as the Alkahest, which I never in any booke of mine ownd my selfe to be a Possessor of, and which I conceive to be so lyable to be misemployed to dangerous purposes, that if I were master of the Alkahest & the way of makeing it, I should thinke my selfe oblig'd not to communicate it to any for whom a long acquaintance had not given me a particular esteeme & friendship[.] And to add that upon the by, I doubt not but Helmonts sayes true where he declares that the Alkahest is tædiosissimae Præparationis; & that other Arcana majora of the Chymical Philosophers, tho the processes were deliverd without Riddles, would be very difficult to prepare by persons not already ‹very› well versd both in the more common, & some of the more subtile operations of Chymistry. Thô I do not therefore blame your aspireing to the highest secrets, yet I am sorry to see no more ground than I do to expect that either by my means or other mens you will easily be made master of them.[39]

It is interesting how Boyle makes a virtue of the fact that he had never made claims to such secrets in his books, as also how he invokes moral scruples about divulging so potentially dangerous a formula (except to those he trusted), and questions the viability of the process. But it is also worth observing that we know from Lawrence Principe's researches that this disavowal on Boyle's part was highly disingenuous, in that he *had* pursued exactly the knowledge which he here effects to disclaim.[40] Indeed, in his later years Boyle made a collection of just the kind of secrets to which Avery evidently aspired, and, though the collection itself is lost, Boyle's prefatory letter to it survives, in which he wrote how:

> since I find myself now grown old; I think it time to comply with my former intentions, to leave a kind of Hermetick Legacy to the Studious Disciples of that Art, & to deliver candidly in the annex'd Paper, some Processes Chymical & Medicinal, that are less simple and plain than those barely Luciferous ones,

39 Boyle, *Correspondence*, vol. 5, pp. 191–3.

40 Principe, *Aspiring Adept*, esp. Ch. 5.

> I have been wont to affect; and of a more difficult & elaborate kind, than those I have hitherto publish'd, and more of kin to the noblest Hermetick Secrets, or as Helmont styles them, *Arcana majora*. Some of these I have made & try'd; Others I have, (thô not without much difficulty) obtain'd by Exchange or Otherwise from those that ‹affirm they› knew them to be real, & were themselves competent Judges; as being some of them Disciples of true Adepts, ‹or› otherwise admitted to their acquaintance & conversation. Most of these Processes are clearly enough deliver'd, and of the rest there is plainly set down without deceitful terms, as much as may serve to make what is literally taught, to be of great utility: thô the full & compleat uses are not mention'd, *partly* because in spite of my Philanthropy, I was ingag'd to secrecy, as to some of these uses, and *partly* because (I must ingeniously confess it) I am not yet, or perhaps ever shall be, acquainted with them my self.[41]

Paradoxically, as death approached, Boyle seems to have aspired to at least a partial openness even in this field, though within this very passage he reverts to such commonplace excuses as the need to protect the secrets of others, or the fact that his knowledge was incomplete.

Earlier, on the other hand, Principe has illustrated how Boyle used standard alchemical techniques of encryption, including word substitution and dispersal, to obscure the meaning of alchemical material that he recorded. A minor element of this appears even in published writings such as *The Usefulness of Natural Philosophy*, though this usually takes the form of withholding crucial information and seeding the text with phrases and formulae which would alert alchemical readers to the significance of the processes that he was expounding.[42] In his manuscripts, however, he was much more overt, not least in using strange word substitutions, as in the following passage in one of his workdiaries which reads in the original as follows:

> With Dicla and Mardium prepare Nigerus by 7 operations, then mix with it Barakid a 12th part, keep it about 3 dayes in digestion till opening the Vessell from time to time toward the latter end, you perceive almost all the Barakid to have gaind the lower part of the Mixture.

(The fifth word from the end is annotated by Boyle: 'quaere, if not the Upper'). What this actually means is:

> With cinnabar and iron filings prepare Mercury by 7 operations, then mix with it Silver a 12th part, keep it about 3 days in digestion till opening the vessel from

41 Hunter and Davis, *Works*, vol. 12, pp. 365–6. For related documents, see Principe, *Aspiring Adept*, pp. 300ff.

42 See Lawrence M. Principe, 'Robert Boyle's Alchemical Secrecy: Codes, Ciphers and Concealments', *Ambix*, 39 (1992): pp. 63–74; Principe, *Aspiring Adept*, pp. 143ff.

> time to time toward the latter end, you perceive almost all the Silver to have gained the lower [upper] part of the Mixture.[43]

Since such substitutions were often used in his laboratory notes, they were evidently partly intended to protect his data from his own assistants, and in this connection an intriguing piece of evidence survives in the form of an oath which Boyle evidently drafted with a view to having his assistants affirm it (interestingly, his own scruples about oaths made him refrain from making them take a vow.) The formula was: 'I do hereby solemnly & faithfully promise & ingage my self that … I wil not knowingly discover to any person whatsoever, whether directly or indirectly, any process, medicine, or other experiment, which he [Boyle] shall injoin me to keep secret & not impart; without his consent first obtain'd to communicate it'.[44] Hence here we do see Boyle invoking secrecy, even if it might be felt that this was legitimate enough in relation to his own personal information management.

Secrecy and Institutionalization

However, perhaps the oddest example of Boyle's ambivalence over secrecy is to be seen in one facet of his relationship with the Royal Society, thus adding to the complexities of understanding the relationship between secrecy and modernizing trends in the science of his day as exemplified by new institutions like the Society. In the 1660s, Boyle had been in the forefront of initiatives to use institutional deposits with the newly founded body as a means by which Fellows with 'a philosophical notion or invention, not yet made out' could have it sealed up 'till it could be perfected, and so brought to light'.[45] The rationale was that 'this might be allowed for the better securing inventions to their authors', in other words the protection of intellectual property rights, and it was evidently in this connection that Boyle availed himself of the facility in the 1660s. Subsequently he reverted to the practice, and his later deposits were opened at a meeting of the Society after his death, on 9 February 1692. Moreover, since the documents involved were described at the time as 'Arcana', there is a clear link to the secrets tradition.[46] The papers in question do indeed turn out to record recipes of just the kind that others might have protected

43 WD 34-55. Cf. Michael Hunter et al., *The Boyle Papers* (Aldershot: Ashgate Publishing, 2007), p. 147.

44 RS MS 189, fol. 13, reproduced in Shapin, *Social History of Truth*, p. 404. See also ibid., p. 403 and Hunter, *Scrupulosity and Science*, p. 67n.

45 Hunter, *Scrupulosity and Science*, pp. 219–20.

46 RS Copy Council Minutes, vol. 2, pp. 113–4, quoted in Hunter and Davis, *Works*, vol. 12, p. xxv. See also ibid., pp. xxv–xxvi.

as 'secrets', but there is a strange contrast between them which encapsulates Boyle's residual ambivalence in this respect.

One paper, dated 30 November 1683 and deposited by Boyle later that year in sealed form, outlined an effective method of testing the saltiness of water using a solution of silver nitrate, a key tool in relation to the patent application for desalinizing sea-water with which he was involved at that point.[47] Interestingly, this was not the only institutional deposit that occurred in connection with the desalinization project. An announcement in the *London Gazette* for 23 January 1685 reveals that various of the recipes it deployed were deposited with the Lord Mayor of London, 'lest a Secret of so great importance to the Publick might come to be lost, if lodged only in the knowledge of a few Persons therein concerned'.[48] It is revealing that the term 'secret' was used in this public statement, though it is significant that the concern was to ensure that the information in question was preserved – a rather Boylean concept, as we have seen. As for the paper deposited with the Royal Society, this was even more 'public' in its rationale, since although deposited in manuscript form, it was in no way an 'arcanum', being presented as ready for print, with elaborate instructions as to how shoulder notes should be placed and the like which were exactly followed when it appeared in *Philosophical Transactions* in 1693.[49]

The second item, deposited on 14 October 1680, however, is rather bizarre.[50] It describes a method of preparing phosphorus from human urine which is evidently the very process which, as we saw earlier, Boyle had discovered by a typical combination of hints from informants and experimentation. What is extraordinary about it is the manner in which it was deposited. It was carefully folded up and secured with three seals, one of them Boyle's own and the others evidently those of two of the four witnesses to the deposit. It really does *feel* like a 'secret' carefully salted away, even if the excitement of the occasion when it

[47] RS Classified Papers, vol. 6, p. 51. For the context see Hunter, *God and Science*, pp. 215ff., esp. p. 218.

[48] Quoted in R.E.W. Maddison, 'Studies in the Life of Robert Boyle F.R.S.: Part II. Salt Water Freshened', *Notes and Records of the Royal Society*, 9 (1952): pp. 196–216 (p. 207). There is no reason to think that Boyle was personally responsible for this (the account of the deposit in the *London Gazette* suggests that it was the King who took the initiative; the process deposited with the Royal Society does not seem to have been included). See also ibid., pp. 204–7. For a different reading of this episode, see Adrian Johns, *Piracy: the Intellectual Property Wars from Gutenberg to Gates* (Chicago, IL: University of Chicago Press, 2009), pp. 73ff.

[49] Hunter and Davis, *Works*, vol. 12, pp. xxviii and 165–74. It is perhaps worth noting that the third paper opened at the same time was a copy of Boyle's queries concerning mineral waters, and it is unlikely that he saw this as an 'arcanum': ibid., p. xxvi, and vol. 10, p. xxx.

[50] RS Classified Papers, 11 (1) 21, dated 30 September 1680. For the deposit, see Hooke's report in Thomas Birch, *The History of the Royal Society of London* (4 vols, London, 1756–7), vol. 4, pp. 55–6.

was opened in 1692 was rather reduced when Boyle's former assistant, Frederick Slare, observed that chemists had since discovered 'a more convenient & shorter way of preparing' the substance (he might have added that the information contained in the paper had in fact been in print since later in 1680, when Boyle had included an almost identical version of 'The Process' at the end of his *Aerial Noctiluca*).[51]

It is perhaps ironic that Boyle arguably came closest to the secrets tradition in the institutional setting where he should have been furthest from it. After all, his deposits with the Royal Society were of details of processes similar to those which their proprietors often refused to disclose, and which Boyle would therefore typically have sought to demystify, as we have seen. Just as it seemed as if – alchemy aside – Boyle displayed a rather modern attitude, he pulls us up short. Indeed, this makes it hardly surprising that to Boyle's successors his attitude towards secrecy often seemed frustratingly inconsistent and unclear. Thus, in reporting to Locke after Boyle's death about their communication on alchemical matters, Newton complained how Boyle had shown a certain 'reservedness' concerning a secret that interested him, suspecting that Boyle had deliberately withheld part of it from him.[52] Similarly, Samuel Cottereaux Duclos of the Parisian Académie des Sciences was critical of Boyle not least for his refusal to divulge key particulars in chemical reactions, which he thought showed an unhelpful double standard on Boyle's part.[53] There seems little doubt that Boyle's equivocation on such matters had a significant effect on his legacy, meaning that it was more mixed than might otherwise have been the case. In all, Boyle seems likely to remain a somewhat paradoxical figure.

---

51 RS Copy Council Minutes, vol. 2, pp. 113–4. Compare Hunter and Davis, *Works*, vol. 9, p. 303, and vol. 12, pp. 163–4 (this overlap is not there commented on: this will be rectified in corrigenda to the edition). See also ibid., vol. 9, p. 300, for Boyle's report on his deposit of this item at the Royal Society. For a commentary, see Partington, 'Early History of Phosphorus', esp. p. 407.

52 See Newton, *Correspondence*, vol. 3, p. 218. See also ibid., pp. 193, 195 and 216–9.

53 See Victor D. Boantza, 'Chymical Philosophy and Boyle's Incongruous Philosophical Chymistry', *International Archives of the History of Ideas*, forthcoming. It is perhaps worth noting here a similar attitude on the part of an early annotater of the 1663 edition of Tome 1 of Boyle's *Usefulness of Natural Philosophy*, evidently Algernon Capel, 2nd Earl of Essex, whose bookplate is in the copy of the book in question, Bodleian Library, Oxford, 12 Θ 1329 (the hand is similar to that of letters from Essex in British Library Additional MS 40629, fols, 199, 209 and 217). He gave vent to his frustration at Boyle's ambivalence over divulging information in a series of sarcastic comments, for instance 'thank you, Sir,' where Boyle wrote: 'what I can further say of this matter, I must not declare in this place' (p. 198 [sic]; *Works*, vol. 3, p. 412 – a typical example of the kind of generalized excuse referred to on p. 95, above). For other annotations see pp. 137–8, 140–41, 146, 170, 175, 177, 181, 183, 185, 188–9, 193, 196 and 354. I am indebted to Will Poole for drawing my attention to this volume.

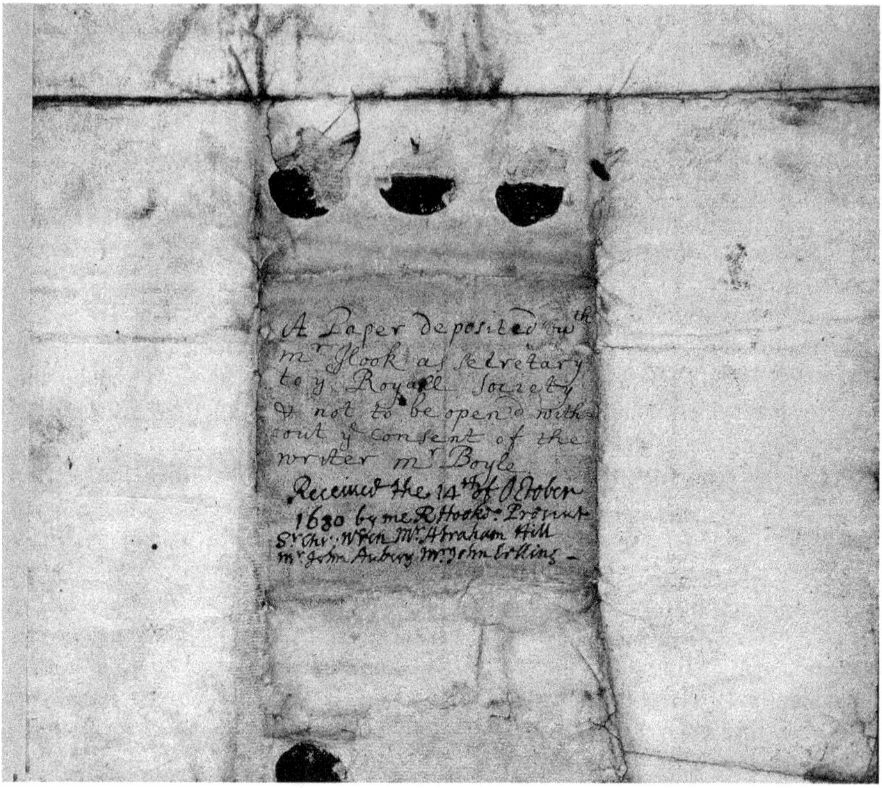

Fig. 4.1 Cover sheet of the document deposited by Boyle at the Royal Society on 14 October 1680 describing the method of making phosphorus from human urine, with an inscription by one of Boyle's amanuenses and an endorsement by Robert Hooke noting his receipt of it and the names of those present, and with the remains of three different seals (Royal Society Classified Papers 11 (1) 21)

Source: © The Royal Society.

Chapter 5
Openness vs. Secrecy in the Hartlib Circle: Revisiting 'Democratic Baconianism' in Interregnum England

Michelle DiMeo*

The members of the so-called 'Hartlib circle', a London-based international correspondence network surrounding Samuel Hartlib, Jan Amos Kaminski (Comenius) and John Dury, comprised one of the most active and intellectually diverse groups dedicated to realizing socio-religious reform c. 1642–60. The 'pact' signed in 1642 by Hartlib, Dury and Comenius promised that their society would dedicate itself 'to the glory of God and the utility of the public'.[1] United by an interest in new scientific discoveries and the idealistic goal of extending all useful knowledge to benefit the commonwealth, the Hartlib circle viewed the dramatic upheaval of the Civil Wars and the utopian potentials of the Interregnum as a political climate ripe with potential for radical change. Their plans for social and educational development were fuelled by a Protestant (but primarily Puritan) impulse to transform England into a utopian godly society, thereby making the Hartlib circle representative of the larger religious, political and scientific climate of Interregnum England.

Partly because of this powerful image of openness which members of the circle projected about themselves, and partly because the Interregnum is still upheld as a time of egalitarian learning and networking prior to the institutionalization of science in the Restoration, scholars have heralded the Hartlib circle as an example of a democratic network dedicated to encouraging open communication in natural philosophy and medicine. For example, the term 'democratic Baconianism' used in the title is a quotation from the last chapter in William Eamon's *Science and the Secrets of Nature*, where he charts the Hartlib circle's goal of changing secrets into public knowledge and shows how this democratic view of learning was not

* Thanks to Sarah Easterby-Smith, Vera Keller and the editors of this collection for their comments on an earlier draft of this chapter.

1 This is taken from the 'pact' signed by Hartlib, Dury and Comenius on 3 March 1642. Quoted in Mark Greengrass, 'Archive Refractions: Hartlib's Papers and the Workings of an Intelligencer' in Michael Hunter (ed.), *Archives of the Scientific Revolution: The Formation and Exchange of Ideas in Seventeenth-Century Europe* (Woodbridge: The Boydell Press, 1998), pp. 35–48 (p. 36).

upheld in the elite climate of Restoration science.[2] This view can be found in many other foundational works on the subject. For example, Charles Webster highlights the subtle differences between how members of the circle viewed particular subjects, such as the relationship between academia and the universal advancement of learning, but he also shows how they married their religious, political and scientific ambitions to produce many proposals throughout the commonwealth aimed at benefiting the public, each with varying degrees of success.[3]

The Hartlib circle did not achieve its ambitious goal of making England into a utopia of openness, and scholars who support the theory of egalitarian Interregnum science point to the Restoration of the monarchy in 1660 as the cause for this failure. Since the Restoration produced a more stable political climate and saw the creation of the Royal Society – an exclusive scientific institution supported by monarchical funds – the goal of openly communicating useful knowledge and the commitment to advancing public services were unable to thrive in the more structured and elitist environment of the Restoration. While the Founder Fellows of the Royal Society professed a commitment to Baconian openness, throughout the 1670s and 1680s the organization became increasingly populated by members less committed to these altruistic beginnings.[4]

Yet the Restoration must not be seen as the sole cause for the failure of Baconian openness in early scientific societies, as ambivalence to incorporating openness is also evidenced in the writings of the most charitable Interregnum reformers. Conflicting branches of secrecy were encouraged within the Hartlib circle itself, and members appear to have held a tacit understanding of when openness or secrecy should be applied. Several letters that passed between members of the circle required the recipient not to disclose to anyone the new knowledge they just received, and secrets were exchanged as a form of currency in a manner similar to that found both before and after the Interregnum. There has been some recent research into the paradox of when secrecy was used in the circle, but these analyses generally result in preserving the harmonious picture of Hartlibian open access and

---

2 William Eamon, *Science and the Secrets of Nature: Books of Secrets in Medieval and Early Modern Culture* (Princeton, NJ: Princeton University Press, 1994), I am using the term 'Baconian' because it is concise and recognizable in scholarship; however, for a more thorough discussion of the Hartlib circle's complex methodological and philosophical influences, see Stephen Clucas, 'In Search of "The True Logick": Methodological Eclecticism among the "Baconian Reformers"' in Mark Greengrass, Michael Leslie and Timothy Raylor (eds), *Samuel Hartlib & Universal Reformation: Studies in Intellectual Communication* (Cambridge: Cambridge University Press, 1994, rep. 2002), pp. 51–74.

3 Charles Webster, *The Great Instauration: Science, Medicine and Reform, 1626–1660* (London: Gerald Duckworth and Co., 1975).

4 Eamon, *Science*, Ch. 10.

free communication.[5] But some of their uses of secrecy cannot, and should not, be coded as indirect paths toward openness; instead, a more helpful line of enquiry would be to explore these contradictions to determine where points of conflict emerge. Stephen Clucas has convincingly argued that 'Hartlib did not dispense with the hermetic injunction of secrecy', and he has demonstrated how Hartlib and Sir Cheney Culpeper, among other members of the circle, maintained secrecy when discussing alchemical matters.[6] However, there has not been a detailed study of how secrecy surfaces in the Hartlib Papers as a whole, nor a survey on the range of topics for which secrecy was required.

This chapter offers an overview of the complicated negotiation of how secrecy was sometimes promoted and required within this larger environment of openness. I will demonstrate that in the Hartlib Papers, the discourse of secrecy arises not only when members consider alchemical matters or when discretion was required by members outside the circle, but also when these correspondents discuss certain political, medical and philosophical subjects amongst themselves. Similar to their conviction in openness, the Hartlib circle's conviction in secrecy appears to have held theological implications, and they were able to reconcile these opposing philosophies by both believing in their own moral superiority and allowing themselves a subjective understanding of when one must adhere to openness or secrecy.

The Ideal of Baconian Openness

In Charles Webster's *The Great Instauration*, which in many ways is still the foundational work on the 'reformation of science' in mid-seventeenth-century England, he explains that Francis Bacon became the most influential scientific authority of the 1640s and 1650s. He suggests that 'Bacon's philosophical system evolved in the context of the Calvinist code of ethics as well as of the providential and millenarian view of history'.[7] Though most of Bacon's key texts were published in the 1620s (and some, such as the *Advancement of Learning* (1605), even earlier), they did not reach their peak in popularity in England until decades after their initial printings, when his suggestions of recapturing a pre-lapsarian state of perfection on Earth found new life in the movements for socio-religious reformation during the Civil Wars and Interregnum.[8] The reformers believed that

5 Gerhard F. Strasser, 'Closed and Open Languages: Samuel Hartlib's Involvement with Cryptology and Universal Languages' in *Samuel Hartlib & Universal Reformation*, pp. 151–61.

6 Stephen Clucas, 'The Correspondence of a XVII-Century "Chymicall Gentleman": Sir Cheney Culpeper and the Chemical Interests of the Hartlib Circle', *Ambix*, 40/3 (1993), pp. 147–70 (p. 148).

7 Webster, *Great Instauration*, p. 25.

8 William T. Lynch, 'A Society of Baconians?: The Collective Development of Bacon's Method in the Royal Society of London' in Julie Robin Solomon and Catherine

the Great Instauration of learning that awaited the public would be coupled with a new understanding of nature itself, as she would reveal her 'secrets'.

One prerequisite to achieving the 'New Atlantis', Bacon's utopian scientific society, required open access to all useful knowledge. The educational reformer Samuel Hartlib employed this philosophy in the early 1640s by organizing an extensive intelligence network and cultivating a reputation as an ambassador for open communication. The nonconformist minister Henry Jessey described Hartlib as one who had a 'most pious & publicke spirit' and was 'searching into the secrets of Nature & communicating [his] discoveries for the Common good', a portrait representative of the reputation Hartlib fashioned for himself.[9] He began collecting information in a centralized location and organizing it to later be distributed more widely for the public good. He hired scriveners, solicited specific information from his international group of talented individuals, and copied and distributed essential print and manuscript texts. In order to reach a more authoritative view of nature, everyone (ranging from accredited physicians, to chemical adepts, merchants and medical lay-practitioners) should openly share their secrets. As Hartlib explained in a letter to the mathematician Adolf Tassius, 'it is indeed to be deplored that the *Eruditi* do not unite better against the *crassam ignorantiam falsam que Scientiam generis humani*, and communicate one to another truly and without deceit whatever each has found true and good in any parts of human science'.[10]

This ideal of freely communicating knowledge is not only evidenced in Hartlib's writings; it was a philosophy central to the circle's larger vision. The commitment to openness permeates letters on various aspects of science written by all key members of the Hartlib circle. One example is a letter concerning agricultural developments written in 1657 by John Beale to an unknown recipient. Here Beale professed his commitment to openness, promising 'Neither doe I hold any of thiese kinds of secrets vpon any other termes then to com[m]unicate them wherever I find they may bee serviceable to o[u]r owne Countrey, or to humane kind'.[11] Beale's letter is characteristic of others in the Hartlib Papers, where correspondents divulged all they knew about a topic and stated their desire for

Gimelli Martin (eds) *Francis Bacon and the Refiguring of Early Modern Thought: Essays to Commemorate The Advancement of Learning (1605–2005)* (Aldershot: Ashgate, 2005) pp. 173–202, esp. pp. 178–81.

9 Letter, Henry Jessey to Hartlib, undated, Hartlib Papers 53/35/3A-4B. The Hartlib Papers are held at the University of Sheffield Library. All quotations from this archive are taken from the Hartlib Papers CD-ROM 2nd edn (Sheffield, 2002). Hereafter abbreviated HP.

10 'It is indeed to be deplored that the Educated do not unite better against the uneducated, ignorant and false Knowledge of humanity'. Letter, Samuel Hartlib to Johann Adolf Tassius, 10 August 1638, British Library (hereafter BL), Sloane MS 417, fol. 213. Quoted in G.H. Turnbull, *Samuel Hartlib: A Sketch of His Life and His Relations to J.A. Comenius* (London: Oxford University Press, 1920), p. 22.

11 Copy Letter, [John Beale] to Unknown Recipient, 8 September [1657?], HP 42/1/34A-35B.

it to be communicated to a larger audience if the recipient deemed it potentially beneficial. The archive also contains copies of printed pamphlets with declarations of public openness, such as Gabriel Plattes's 'The Profitable Intelligencer' (1644) and Hartlib's 'The Reformed Virginian Silk-Worm' (1652). Further, the very act of writing and sending a letter to Hartlib must be viewed as a public activity with an aim towards advancing common knowledge, as correspondents knew that he might copy, annotate and distribute their letters if he thought they contained useful information – something he occasionally did before receiving the author's permission.[12] If there was a specific reason why the contents of a letter should not be made public, the correspondents knew they should include a note explaining their reasons for discretion and clearly asking Hartlib to keep this information private.[13]

Since everyone was morally obliged to participate in openness in order to benefit society at large, those who harboured secrets were unequivocally admonished by members of the Hartlib circle. In a letter to Hartlib in 1659, Henry Oldenburg condemned a 'Monsieur Borreel' for being 'avaricious of his secret'.[14] When one of Hartlib's associates, an 'R. Jones', heard in 1660 that the chemist Johann Rudolph Glauber was not sharing his new 'secret … of fertilising lands', Jones lamented, 'it is pitty, he should not be sent into Arabia or Lybia, and the like quarters for to practise his skill and enrich the world'.[15] Lady Katherine Ranelagh was angered that 'Monsieur Fountain' would not share his medical recipe for the stone, explaining to Hartlib that 'Hee is of the Common opinion of loving more to keepe a Secret, then to doe good by publishing it'.[16] Her dedication to openness led her to further pursue him. When she finally received the recipe for this medicinal water, which Monsieur Fountain told her he valued 'as a rare secret', she adhered to the circle's oath of openness and shared it with Hartlib.[17] Members of the

[12] For example, Hartlib printed some of Dorothy Moore's letters to Lady Ranelagh in the anonymous pamphlet 'Madam, although my former freedom' (London, 1645) without gaining Moore's permission. For Moore's angry response, see Letter, Dorothy Moore to Hartlib, undated. HP 3/2/143A-144B.

[13] For example, see Letter, John Dury to Hartlib, 12/22 October 1643, HP 3/1/9A-10B. The recipient of this letter is identified as Hartlib in G.H. Turnbull, *Hartlib, Dury and Comenius: Gleanings from Hartlib's Papers* (Liverpool: University Press of Liverpool, 1947), p. 235.

[14] Letter, Oldenburg to Hartlib, 30 April 1659, HP 39/3/22A-23B. Probably the French natural philosopher Pierre Borel (c. 1620–71), whom Oldenburg had written to four days earlier.

[15] Letter, R. Jones to Hartlib, 24 January 1660, HP 44/9/2A-2B. This may be Richard Jones, Lady Ranelagh's son, who was tutored by Hartlib circle members Henry Oldenburg and John Milton.

[16] Extract from Letter, Lady Ranelagh to Hartlib, 10 February 1657, HP 60/13A-14B. Possibly Edward Fountain. See 'Introduction', p. 2.

[17] Extract from Letter, Monsieur Fountain to Lady Ranelagh, undated, HP 60/4/201A-202B.

Hartlib circle viewed those who did not cooperate with their call for openness as individuals that placed personal gain before the public good; therefore, they were both selfish and morally weak. Hartlib and his correspondents were confident enough in their larger social ambitions and their personal commitments to God that they actively pursued, and remained frustrated by, outsiders who maintained secrets.

Because they promoted open communication as a means to achieving their ambitious plans for social reform, the Hartlib circle has become representative of the democratic scientific environment of Interregnum England. However, while this is an image that Hartlib and his associates projected about themselves, they also had a preoccupation with secrecy that is only thinly veiled in their manuscript correspondence and print publications. But when was secrecy permitted, or even required, in this utopian plan for openness? How did members of the circle know when to keep secret, and why were they not judged in the same way that they judged others who harboured secrets?

Secrecy in the Hartlib Circle

Though Hartlib's philosophical and theological sources promoted open communication as a means to achieving a utilitarian advancement of learning, many of these documents also reserved a space for necessary secrecy. Jerry Weinberger has illuminated how Bacon advocated secrecy in many of his key works responsible for the rise in 'Baconian openness'. For example, in *The Advancement of Learning*, Bacon promoted the ancient method of selecting material for open distribution 'with judgment and discretion' so that privileged knowledge could not be abused by the common people.[18] Another example is Bacon's posthumously published utopian fiction *New Atlantis*, which revealed the previously secret scientific society Bensalem. The Father of Salomon's House explained that this society selected what knowledge should be made public and 'take all an oath of secrecy, for the concealing of those which we think fit to keep secret'.[19] Though Bacon's works are best remembered for their promotion of freely communicating all knowledge, in reality the information chosen for public dissemination was carefully selected.

Within this context, it is not surprising to find that there were several matters discussed in letters exchanged between Hartlib and his correspondents which required secrecy. Despite the numerous examples of Hartlib circle members

---

[18] Jerry Weinberger, 'Francis Bacon and the Unity of Knowledge: Reason and Revelation' in *Francis Bacon and the Refiguring of Early Modern Thought*, pp. 109–27 (p. 115).

[19] Francis Bacon, *New Atlantis and The Great Instauration*, ed. Jerry Weinberger (Arlington Heights, IL: Harlan Davidson, 1989), p. 82. Also see Weinberger's introduction to the volume, esp. pp. xxx–ii.

professing their commitment to openness and disparaging people who harboured secrets, many of their letters also demonstrate that they shared privileged knowledge with each other under a vow of secrecy. Some scholars have noticed this trend and have suggested that the reformers were sometimes bound to secrecy by external means beyond their control. Yet these letters do not always include a reason or apology for requiring secrecy; instead, they operate under the assumption that both writer and recipient share a tacit understanding that not all secrets should be made public.[20] Throughout the two decades of their recorded correspondence, many letters that John Dury wrote to Hartlib included a phrase such as 'keep it to yowr self as a secret', often with little or no explanation of why this secrecy was necessary.[21] A letter that the English diplomat Sir Thomas Rowe sent to John Dury, relaying information on foreign politics, concluded with the caveat, 'These th[ing]s I write are very secret knowen to few and must not bee made Comon-Newes, but only intimated to the L[or]d Chancelor'.[22] Benjamin Worsley, who was heavily involved in many of Hartlib's most ambitious projects for social reformation and for extending access to useful information, was so secretive of his knowledge that John Moore wrote to Hartlib, 'I thought you had better known, then to think he [Worsley] would impart his secrets to any one'.[23] Though Lady Ranelagh was critical of Monsieur Fountain for withholding his medical secrets, Henry Oldenburg referred to her as 'a person that can keep a secret as well, as any I know', suggesting that she maintained a contradictory reputation as one who promoted open access to all useful information, but also as one who knew when and how to 'keep a secret'.[24]

In reality, the Hartlib circle was a diverse network comprised of many people who did not necessarily know each other, but were connected through a few common nodal points.[25] Therefore, sub-circles formed within the larger circle, in which members shared their secrets with confidants whom they knew to be trustworthy.

---

20 Eamon, *Science*, Ch. 10.

21 Letter, John Dury to Hartlib, 8 September/29 August 1644, HP 3/2/57A-B. For further examples, see Letter, John Dury to Hartlib, 1646, HP 3/3/21A-B and Letter, John Dury to [Hartlib], 24 May/3 June 1661, HP 4/4/20A-21B.

22 Copy Letter, [Sir Thomas Rowe] to [John Dury], [18?] July 1636, HP 14/4/45A-46B.

23 Letter, John Moore to Hartlib, 28 July 1645, HP 21/8/3A-B. For more on Worsley, see Charles Webster, 'Benjamin Worsley: Engineering for Universal Reform from the Invisible College to the Navigation Act' in *Samuel Hartlib & Universal Reformation*, pp. 213–35.

24 Letter, Henry Oldenburg to Hartlib, 5 July 1659, HP 39/3/25A-27B.

25 Timothy Raylor, 'Newcastle's Ghosts: Robert Payne, Ben Jonson, and the "Cavendish Circle"' in Claude J. Summers and Ted-Larry Pebworth (eds), *Literary Circles and Cultural Communities in Renaissance England* (Columbia, MO: University of Missouri Press, 2000), pp. 92–114 (pp. 92–3); David Lux and Harold Cook, 'Closed Circles or Open Networks?: Communicating at a Distance During the Scientific Revolution', *History of Science*, 36/112 (1998), pp. 179–211.

Correspondents chose who could be in their sub-circles through a process of self-selecting and self-validating each other, based on an individual's intellectual and moral reputation. The motive behind their decision to maintain secrecy is different in each situation, but the reasons include those that were religious, economical, personal and political. The secrets exchanged within the Hartlib circle also ranged, including political news, medical recipes and agricultural discoveries. Like most contemporaries, they usually did not articulate a clear distinction between various types of secrets, as a secret of empire could also be a secret of nature.[26]

A preliminary search through the Hartlib Papers produced over 100 direct references to members of the circle maintaining or promoting secrecy between each other.[27] Though some of these situations show correspondents bound to secrecy because of external causes, such as political safety or having received a recipe under the promise of not disclosing it, the majority of these clandestine exchanges operated under the assumption that secrecy was desirable for reasons freely chosen by sender and recipient.[28] Most displays of secrecy in the Hartlib circle appear to have been because of one of three reasons. First, an ability to keep a secret was sometimes necessary to establish or preserve one's reputation. Second, secrets could function as intellectual currency, or gifts to be exchanged in a sub-circle of friends. Third, since many members of the public were of ill moral repute, members of the Hartlib circle made it their responsibility to guard from others information they deemed potentially dangerous.

Secrecy to Maintain One's Reputation

Since the Hartlib circle involved international exchanges between a diverse range of new and old correspondents, those who participated had to establish their reputation. This could be done in different ways: sometimes individuals were recommended to Hartlib by a mutual friend, or sometimes Hartlib solicited information from trusted sources about an individual about whom he had heard. When drawing new information from an international pool of unknown projectors,

---

[26] Vera Keller, 'Mining Tacitus: Secrets of Empire, Nature, and Art in the Reason of State', forthcoming in *BJHS* special issue, Daniel Margocsy and Koen Vermeer (eds).

[27] This is based on my keyword search of the CD-ROM. While a search for words like 'secret' and 'secrecy' predictably produced hundreds of hits, I treated each hit individually and found over 100 direct references to promoting and maintaining secrecy.

[28] In a letter from Joseph Avery to his brother Samuel, he explains that he cannot reveal his political secrets because it 'would expose all that I haue besids vnto hazard, and danger'. Letter, Joseph Avery to Samuel Avery, 16 June 1642, HP 45/3/15A-17B. For an example of secrets being exchanged under a promise of secrecy, see 'Ephemerides, 1658', HP 29/7/16A: 'Mr Boyle know's a Friend that hase the R[ecei]p[t] of the Hamburg Philosopher for Vniversal Husbandry. Hee makes his Apples to grow very big etc. but Mr Boyle is bound to secrecy and binds others to it'.

medical lay-practitioners and chemists, the individual's reputation was an important first step in determining the validity of their project and their reliability as a partner.[29] Since anyone could be a fraud or an avaricious cheat, one way to establish oneself as a person of intellect and civility in a new community was to demonstrate the ability to keep a secret. In this sense, secrecy functioned as evidence of one's trustworthiness.

In many ways, the Hartlib circle followed the codes for identifying as a virtuoso established in England in the early decades of the seventeenth century. William Eamon has shown that the virtuosi developed this identity in contrast to both scholars and the vulgar sort, linking curiosity and experimental natural philosophy with civility and genteel manners.[30] By demonstrating an interest in and knowledge of nature's secrets, as well as an ability to maintain secrecy, one could begin to establish one's reputation as a virtuoso. However, while Eamon suggests that the Hartlib circle moved away from this more conservative scientific climate of Jacobean and Caroline England by transforming secrets into public knowledge, the discourses within the Hartlib circle sound surprisingly similar, viewing the preservation of secrecy as a distinguishing factor between those of intellectual and moral repute, and frauds. Since the language of self-identifying as a virtuoso in England was relatively new, Interregnum virtuosi relied on the few rules in place, even if they may have appeared antithetical to the current social and intellectual environment of openness.

Though Hartlib and his closest associates were vocal about and known for their shared commitment to enhancing public knowledge, the circle still obeyed the main dictates of virtuosity and displayed their ability to keep a secret as proof of reliability and trustworthiness. When Stephen Bradwell wrote his first letter of introduction to Hartlib, he professed, 'I pray you good Sr doubt not of my secrecy in any of your business; neither of my sonnes. For we see that your Inventions & Endeavours aime at the Good of Christs Church & we professing the same: dare not make a shew of Godlinese & deny the power of it'.[31] Strikingly, Bradwell recognized Hartlib's larger socio-religious goal for open access to all useful knowledge, but still he used the traditional language of the virtuosi to signal his cooperation with the project. A similar sentiment was expressed by Francis Sanderson to Hartlib when the two began communicating again after a lengthy period without contact. Sanderson began his letter to Hartlib saying that he hoped 'you will the freelier communicate your thoughts to mee: w[hi]ch I doe obleidg my selfe shalbe most sacredly kept secret'.[32] In addition to using secrecy as a way to introduce oneself as a respectable and trustworthy source, a promise of

29 Lux and Cook, 'Closed Circles'.

30 Eamon, *Science*, Ch. 9.

31 Letter, Stephen Bradwell to Hartlib, 12 January 1644, New Style Calendar, HP3/14/3B. Stephen Bradwell may be the London physician of the same name.

32 Letter, Francis Sanderson to Hartlib, 16 January 1659, New Style Calendar, HP9/2/1A-B.

secrecy could be used in an established friendship to signal one's devotion to a new project. When Hartlib asked the mathematician Robert Wood to investigate the papers of, and to work with, William Potter, Wood expressed his willingness to participate with a proclamation towards secrecy: 'I am most willing & ready to comply wth your desires of engaging my selfe seriously in this affaire … engaging my[self] to ye greatest secrecy'.[33]

While the ability to maintain a secret was the mark of a virtuoso, indiscrete openness was considered vulgar. In 1642, Joseph Avery, a Royalist exile in Hamburg, wrote to his London merchant brother Samuel Avery about needing to prove his innocence after suffering from political slander. Though communicating information that he received from Charles I 'would soone cleare all', Avery knew that he must keep secret partly because 'the revealing of such secrets would … wound my reputation'.[34] Similarly, the explicit search for another's secrets could signal greed or unfamiliarity with gentlemanly conventions, and therefore damage one's reputation. Since Hartlib often asked his friends to gain information about the latest secrets about which he had heard, these individuals were placed in the difficult position of needing to gain specific information from a source without appearing as a disrespectful and untrustworthy charlatan. Robert Wood's process of analyzing Potter's papers for Hartlib demonstrates one such complicated situation: Wood promised Hartlib that he would discern Potter's method – a commitment which required him to display his reputation as a virtuoso by demonstrating he could maintain secrecy – but he could not appear to Potter as one who was too desperate to gain his secret, as this could make him appear deceitful. As Wood explained in a letter to Hartlib, 'I do not desire the discovery of his designe, because he may perhaps looke upon it as a secret'.[35] Virtuosi may allow other virtuosi to keep secrets, as it signalled an awareness of the occasional requirement of secrecy to which they were all bound. By appearing too eager to learn of Potter's secret, Wood would damage his reputation as a virtuoso, appearing as one who does not understand these tacit codes of conduct.

The paradox in these examples is that an ability to keep a secret signals a higher moral and social status, and therefore demonstrates one's commitment to Hartlib's goal of public openness. Steven Shapin argued that an individual's reputation (determined by moral integrity and genteel civility) was important to seventeenth-century scientists when determining the credibility of experiments.[36] While Hartlib appears to have accepted information from a wide range of sources with little discretion, he was careful to consider an individual's reputation before regarding them as an active correspondent in the circle. By offering a vow of secrecy, and thereby promoting themselves as men and women of integrity,

33 Letter, Robert Wood to Hartlib, 16 November 1658, HP33/1/33A-35B.

34 Letter, Joseph Avery to Samuel Avery, 16 June 1642, HP45/3/15A-17B.

35 Letter, Robert Wood to Hartlib, 21 March 1658, HP 33/1/48A-49B.

36 Steven Shapin, *A Social History of Truth: Civility and Science in Seventeenth-Century England* (Chicago, IL: University of Chicago Press, 1994).

Hartlib's associates may signal their worthiness to participate in the circle's many projects and disassociate themselves from the vulgar. In the case of evaluating reputation, indiscretionary openness was suggestive of low moral values, which could possibly result in exploiting knowledge for personal gain. A true commitment to public openness must be exhibited by an ability to withhold private knowledge, an act which ultimately places social benefits above private interests.

Secrets as a Form of Currency

The second way that secrecy was used by the Hartlib circle was to ensure the value of certain information: by preserving the exclusiveness of a secret, it may serve as a form of currency. Craig Muldrew has demonstrated how words functioned as a form of capital in the early modern 'market of communication'.[37] Because credit was dependent upon an individual's reputation, words could have a substantial economic impact on people, particularly among the middling sort.[38] Though words did not have a constant market value, such as gold and silver, they were exchanged among people for other words, items, or even money. In Hartlib's 'Ephemerides' of 1656, he relays the story of a 'Mr. Hile or Hoyle' who knew of 'a publick vseful Invention of turning Iron into steele wch may prove very beneficial to the Revenues of the Publique'. However, Hartlib notes that this man holds it as a 'secret hee tenders to reveale to his High[nes]s vpon condition to bee provided for with an Annuity for himself and children'.[39] In this situation, it is by maintaining secrecy that these words can be exchanged for money. If 'Mr. Hile or Hoyle' publicly disclosed his secret, it would lose its value.

While selling secrets for money was discouraged by the Hartlib circle, correspondents exchanged secrets amongst themselves, signalling that such information functioned as a type of currency in the circle, though not one of monetary value. As a recognized form of intellectual currency, one may choose to harbour a secret until there is one of equal value for which one would like to exchange it. It is the very rarity and exclusiveness of these items that ensured their value, so the currency was dependent upon maintaining secrecy. One anonymous extract in a letter to Hartlib concerning Johann Sibertus Küffler's portable oven offers a perfect summary of this: 'if one should goe to explaine it, then would it noe longer be an Art or Secret, neither is it also soe laudable'.[40] A secret must be kept 'a secret' in order for it to maintain its value.

---

[37] Craig Muldrew, *The Economy of Obligation: The Culture of Credit and Social Relations in Early Modern England* (Basingstoke: Palgrave Macmillan, 1998), p. 138.

[38] Ibid., Chs. 5 and 6.

[39] 'Ephemerides' 1656, Part 3, HP 29/5/88A.

[40] Extracts on Küffler's *Portable Oven & Invention for Fresh Water*, undated, HP 55/9/3A-4B.

Several examples of correspondents exchanging secrets with each other may be found in the Hartlib Papers. Benjamin Worsley explained in a letter to John Dury that he received William Petty's 'Art of Brewing in wooden-vessels … by exchanging some secret with him'.[41] A few months after William Potter and Robert Wood began working together, Potter responded to Wood's previous letter by outlining a transaction of secrets, saying, 'yow desire not to be partaker of any such secret of mine wthout imparting yr owne at ye same instant, w[hi]ch I cannot but acknowledge to be exceeding rationall & faire'.[42] The two thereby negotiated a scheme of 'mutuall disclosing of secrets', which both solidified their working relationship and ensured that each party had reciprocal gain. In his 'Ephemerides' of 1655, Hartlib noted that Robert Boyle heard about a new husbandry secret that one could attain by exchanging another secret: 'Mr Boyle know's one that hath a very rare and singular experimented way to preserve naturally all manner of Fruits w[hi]ch is no small Secret in Husbandry. And this may bee by way of exchange communicated to obtain the philosophical Water or some other great Secret'.[43] As this passage suggests, some secrets were worth more than others, and each member of the transaction had to agree that the secrets exchanged had equal value.

As with all commodities, secrets could take the form of gifts to be given to friends. Eamon highlights a process of gift exchange between Oldenburg, Hartlib, Robert Boyle and Lady Ranelagh in 1659, when all four were part of the Hartlib circle and had known each other for many years.[44] Oldenburg sent Hartlib a recipe for a 'Chymicall process of vitriol (in acknowledgem[en]t of ye secret you sent me, w[hi]ch shall not loose the name of a secret for me)'.[45] He then asked Hartlib to 'communicate it to none, but noble Mr Boyle, who, I am sure, upon my desire will impart it to none but MyLady Ranalaugh, w[hi]ch is a person, yat can keep a secret as well, as any I know'. This friendly exchange of secrets is interesting particularly because of the individuals involved. Earlier examples in this chapter have shown Oldenburg and Lady Ranelagh to be committed to openly communicating secrets and critical of those who did not comply with their policy of openness. However, this exchange evidences a history of transacting secrets among this selective group of friends, demonstrating their ambivalent manner of determining when secrecy should be maintained. A similar situation may be found in the correspondence of 1657, the year in which Beale wrote Hartlib the aforementioned letter declaring his commitment to openness. In this same year, he wrote another letter to Hartlib which offers insight into how secrets still functioned as currency in the circle and how those reformers who were vocal about their commitment to openness still maintained some secrets. After telling Hartlib of a new way to brew beer using heath instead of hops, which had considerable health and cost benefits, Beale

41 Copy Letter, Benjamin Worsley to [John Dury], undated, HP 26/33/7A-8B.

42 Copy Letter, William Potter to [Robert Wood], 15 March 1659, HP 30/3/11A-12B.

43 'Ephemerides' 1655, Part 4, HP 29/5/52B.

44 Eamon, *Science*, pp. 344–5.

45 Letter, Henry Oldenburg to Hartlib, with enclosure, 5 July 1659, HP 39/3/25A-27B.

concluded with, 'If yu find it true newes yu may exchange it amongst friends for love, amongst Mysterious men for philosophical Secrets. If yu set it off too cheape, & too hastily, it will fall into contempt'.[46] This concise statement summarizes the multiple ways in which secrets functioned as commodities in the Hartlib circle. Secrets may be given as gifts to friends or exchanged among virtuosi to gain other secrets. Beale's final clause highlights how these secrets may lose value if they are given away 'too cheape', and his implication that it could 'fall into contempt' is suggestive of the final motive for maintaining secrecy: moral implications.

Secrecy for Moral Reasons

Throughout the Hartlib Papers, there is ample evidence that members of the Hartlib circle shared a general distrust of the public and a belief that they were morally superior to the common people. This paternalistic view of society resulted in a system of filtering what information should be publicly distributed. Their belief in their own moral superiority was an extension of the Baconian philosophy that God was revealing his secrets of nature specifically to them. As Hartlib noted in his 1639 'Ephemerides', 'The greatest philosophers should addresse themselves more to God in prayers and in a holy life and so they should finde out more the secrets of Nature then ever they have done'.[47] Many members extrapolated this theory to suggest that if God chose to reveal his secrets to them, this meant they were selected by God because of their religious devotion and moral integrity; therefore, they were responsible for deciding if this new knowledge should be made public or kept secret. As Cyprian Kinner once said to Hartlib, 'To know too much can sometimes be to know unwisely'.[48] John Dury applied this philosophy to suggest that his new knowledge should be 'kept secretly in a pavilion from two great evills, from ye pride of men & from ye strife of tongues'.[49] This discrimination on moral grounds allowed the members to use the language of openness and secrecy in selective ways, granting themselves the authority to determine if the public was ready for the new knowledge they found. Because there were enough corrupt people in society who would use God's secrets for private gain or to advance their malicious plans, there were some situations where keeping such knowledge secret might be a greater benefit to the public.

The moral responsibility to withhold some secrets from the public was a particular concern for money-related matters, such as reproducing coinage. In Pierre Blondeau's memorandum proposing a new way for the government to mint coins, a copy of which is held in the Hartlib Papers and includes Hartlib's manuscript annotations, Blondeau answers concerns about the public counterfeiting money

46 Letter, John Beale to [Hartlib?], 27 April 1657, HP 52/9A-10B.

47 'Ephemerides' 1639, Part 3, HP 30/4/26B.

48 Letter, Cyprian Kinner to Hartlib, 1646, HP 1/33/39A-42B.

49 Copy Letter, [John Dury?] to [Hartlib?], 21 July 1655, HP 4/3/111A-113B.

with a promise of state secrecy: 'whereas some men doe object that assone as the way I use shalbee made publick, then it may bee counterfeited. To that I answer first that the Invention needs not bee made publick; but if it bee the pleasure of the State, the Markins wherewith the brims are marked may bee kept secret among few men, who shalbee sworne to keepe it and not to reveale it to any'.[50] Because some members of the public might use this secret to counterfeit money, those 'few men' privileged by receiving it should protect public interest by maintaining secrecy.

Recipes for alchemical production of the philosopher's stone, which purported to transmute base metals into gold, were viewed with similar caution. If everyone knew how to make gold, the value of it would crash and leave the economy in ruins; therefore, only morally sound individuals should have access to such information. One alchemical extract in the Hartlib Papers notes that it 'was kept so secret yat the name might not bee knowne least it might haue come into the hands of wicked men, & so much harme might haue come thereby'. In addition to practical economic concerns, alchemical information also had to be guarded from the vulgar because many saw it as divine knowledge to be passed only to a chosen few. The extract continues: 'this holie science wch god hath giuen to such as Loue and serue him should if wicked persons had got it beene to them a stramth in their wickedness'.[51] While someone who 'Loue[d] and serue[d]' God would use the philosopher's stone to gain a better understanding of God's world and achieve a higher state of spiritual purity, a 'wicked [man]' would use this power for evil purposes, such as counterfeiting gold or charging exorbitant amounts of money for access to the stone's purported health benefits. Since alchemy was a 'holie science' which God revealed only to those whom he carefully chose, knowledge of the subject should be maintained within a self-selected circle of morally and socially respectable individuals, and kept away from such 'wicked persons' who might abuse the power.

The circle's moral discretion extended to secrets not directly related to reproducing money, including those that could be used by greedy individuals for great monetary gain. Hartlib collected stories of such people in his 'Ephemerides'. One example is a story Robert Boyle told him about a woman who had a cure for a plague that struck Ireland and tried fleeing Dublin without revealing the secret, and another example comes from Robert Child, who told Hartlib about a man in Amsterdam who was robbed of his money and secrets.[52] With so many immoral people in society, individuals must be careful not to reveal their secrets to others who may use them for private gain. It is therefore not surprising to find a note in Hartlib's 'Ephemerides' of 1649 where he is cautious of who should know about Cressy Dymock's new

50 Copy Memorandum on Coinage, Blondeau, undated, HP 39/2/118A-121B.

51 Alchemical Extracts in Scribal Hand, HP 55/16/1A-14B.

52 'Ephemerides' 1655, Part 2, HP 29/5/16A; 'Ephemerides' 1653, Part 3, HP 28/2/66B.

development in husbandry. After recounting the positive results of Dymock's recent experiments with tobacco, Hartlib notes, 'This Experimt may prove a suddain Meanes to grow rich. But it must bee kept mightly secret, otherwise the venting or markets of it may bee blasted by malitious mouthes by aspersing of it that it is sophisticated'.[53] While the Hartlib circle held utopian hopes and millenarian beliefs, their desire for a radically new social order was selective. For as public and egalitarian as their claims to advancing knowledge were, they recognized there were those who 'may make euill use of it'.[54]

Though this self-righteousness had Calvinist overtones of predestination, suggesting God had chosen them as his disciples of knowledge, the implications of moral superiority did not go unquestioned by Hartlib and his discerning correspondents. Within the archive is a series of printed pages with marginalia in Hartlib's hand which addressed the possibility of reconciling personal feelings of moral righteousness with the circle's larger professions to help the public. A very pointed criticism arises in the text: 'are we of all Saints the onely Abrahams from whom God will not hide his secrets? brethren do not think you have no need of more humility'.[55] These doubts are also voiced in some letters exchanged between the correspondents. A letter from Comenius to Hartlib in 1646 states, 'But I am not firmly commited to the last idea (about keeping these major works secret and hiding them from the world until the Lord has given a sign)'.[56] While secrecy was used in a selective manner by members of the Hartlib circle, they were still committed to their personal religious beliefs and to the circle's commitment to socio-religious reform. Promoting secrecy resulted in a complex process of self-reflection, where members considered the fluid distinction between the contradictory codes of conduct they promoted.

The moral responsibility to maintain secrecy may be best illustrated by analyzing the Hartlib circle's most famous project. In 1647 and 1648, Hartlib implemented some of Bacon's organizational strategies as presented in *Novum organum* (1620), and he borrowed heavily from Théophraste Renaudot's proposal for the *Bureau d'adresse*, when he and his associates prepared proposals for their 'Office of Publicke Address'. The first public proposal was issued by John Dury in 1647, and it included the ambitious goal 'all that which is good and desirable in a whole Kingdome may be by this means Communicated unto any one that stands in need thereof'.[57] The Office would comprise two components: the Office of Address for Accommodations (which would facilitate a labour exchange similar to that in Renaudot's *Bureau*) and the Office of Address for Communications (which would collect, organize and distribute to the public all

[53] 'Ephemerides' 1649, Part 1, HP 28/1/7A.

[54] Letter, John Dury to Hartlib, 12/22 October 1643, HP 3/1/9A-11B.

[55] HP 39/2/113/1A-2B.

[56] Copy Letter, Comenius to Hartlib, 25 May 1646, HP 7/73/1A-6B.

[57] John Dury, 'Considerations tending to the Happy Accomplishment of Englands Reformation in Church and State' (1647), p. 41.

useful information). The means by which their goal of expanding communication would be accomplished was explained further in a pamphlet printed in 1648, which said, 'all Things necessary, profitable, rare, and commendable, which are extant in several Places, and scattered here and there, are brought together; and exposed to the View of every one that shall be willing to see them', resulting in such information being 'more useful unto the Publick a Hundred Fold'.[58] However, the functioning of this Office depended on the public trusting in the organizers' professed commitment to enhancing the public good: 'For most Men will not intend any Publick Aim till they can secure their own Interests, and see a Way to get Advantage by that which they call the Publick: But we shall never aim at this; our Delight shall be, that all may be advantaged, and the Publick Interest of the Commonwealth settled, although it should be to our Cost and Disadvantage'.[59] From the beginning of this promotional pamphlet, it is clear that the organizers of the Office declared themselves morally superior to the majority of the commonwealth; those who participated would have to trust in the organizers' commitment to the public as they disclose their knowledge of any given subject (which would range from notifying the Office of all births, deaths and marriages to revealing trade secrets). After people openly communicated to the Office what they knew, these registers 'must be again and again subdivided, and especially that some must be kept secret, and some exposed to the common View of all'.[60] The secret register would include personal data, such as people's addresses, and other information would be withheld from the public but forwarded to the government.[61]

Like all Hartlib's projects, the Office was committed to the public good, and it professed an obligation to advance common knowledge. However, just as the letters and printed texts in the Hartlib Papers demonstrate that secrecy was selectively used and enforced by the circle, the organizational structure of Hartlib's ambitious association for open communication also required secrecy. The 'Office of Publick Address' worked under the assumption that those involved in the planning of the Office, and those who would run its services, were morally superior to the general public. Because some people could use information for evil purposes, the Office would be responsible for filtering new knowledge before it could be made public. Similar to the hermetic injunction that alchemical knowledge should remain with a privileged few, Hartlib's Office would select what information the public could be trusted with knowing. While in one sense this is another example of limited secrecy being required in order to maximize later potential public benefits, the

58 'A further Discovery of the Office of Publick Address, &c.' (London, 1648) in *Harleian Miscellany*, vol. 6 (London, 1745), pp. 13–26 (p. 24).

59 Ibid., p. 15.

60 Ibid., p. 25.

61 Ibid., pp. 24–5; Justin Stagl, *A History of Curiosity: The Theory of Travel 1550–1800* (Chur: Routledge, 1995), pp. 139–43.

moral hierarchy and the distrust of the public sit uneasily against the backdrop of Hartlib's proposed society for openness.

Conclusion

When these three categories are viewed together, the Hartlib circle appears to have suffered from many of the same social biases found in learned societies both before and after the Interregnum. While the Hartlib circle professed a commitment to expanding knowledge by providing open access to all useful information, they still incorporated secrecy into their daily correspondence and their larger public schemes. Members not only maintained secrecy to preserve their reputations and the value of certain information, but also held a critical view of society which allowed them to selectively filter information before making it public. Since the circle comprised people of high moral repute, they elected themselves as those responsible for determining who among the public reached their standards of social and moral responsibility. Therefore, the Hartlib circle and its objectives cannot be considered as egalitarian as they claim.

Hartlib's larger commitment to the public good should not be undermined by these discoveries, but the complicated negotiation between openness and secrecy may suggest that the larger environment of 'democratic Baconianism' in Interregnum England was perhaps not so democratic. Secrecy was used by virtuosi in intellectual exchanges throughout the 1640s and 1650s, and in many ways Hartlib and his associates were a product of their time.[62] Hierarchies in intellectual societies, both formal and informal, were also present prior to the Restoration, and may be evidenced in the utopian ambitions of the Hartlib circle. While this circle was more egalitarian than the Royal Society, and while the relationship between science and society was viewed more radically in the Interregnum than in the Restoration, this chapter suggests that perhaps some of these differences have been exaggerated. Recent research into Robert Boyle has exposed him to be a more complicated man than was previously supposed, and future research on Samuel Hartlib and his associates may offer similar complexities.[63] Instead of remembering the Hartlib circle by its self-fashioned image of democratic Baconians, we may learn more about the circle if we explore the subtle contradictions offered in the correspondents' extant papers, thereby providing a more nuanced understanding of the premier intellectual society in mid-seventeenth-century England.

[62] For an Interregnum example of secrecy outside the Hartlib Circle, see Letter, the Earl of Conway to Edward Harley, 21 October 1651: 'Coronell Grantham gave me somme of the Powder to stay a Flux, he would not tell what it was, by any entreaty, but I have found it out'. BL, Additional MS 70006, fol. 239. I must thank Bernard Capp for this reference.

[63] Lawrence Principe, *The Aspiring Adept: Robert Boyle and his Alchemical Quest* (Princeton, NJ: Princeton University Press, 1998). Also see Michael Hunter's chapter in this volume.

PART III
Illicit Secrets

Chapter 6
Anna Zieglerin's Alchemical Revelations

Tara Nummedal*

In the sixteenth century, central European alchemists played on the full range of meanings attributed to secrecy as they offered their skills and knowledge to patrons and other practitioners. Many alchemists underscored the economic potential of their secrets. Some touted new metallurgical techniques, for example, meant to appeal to princes and bankers with entrepreneurial interests in large-scale mining enterprises. This kind of secret, which could boost one mining or smelting centre's output and bring substantial profits, could take on the aura of state secret as a way of preventing it from falling into the hands of political or economic rivals. Other alchemists proffered smaller scale (if equally valuable) secrets, drawn from newly discovered texts or contact with mysterious adepts. All of these kinds of secrets added value in an increasingly vibrant marketplace for alchemical expertise in early modern central Europe.[1]

The commodification of knowledge fuelled by factors such as the book trade, entrepreneurial aspirations of princes, merchant networks and the new imperial ambitions of globalizing states is not the only explanation for the success of alchemical secrets in the sixteenth century, however.[2] Alchemy was also associated with esoteric and divine wisdom, and this connection sustained its appeal to those seeking privileged insights into the cosmos throughout the early modern period. There were a variety of contexts for this kind of divine alchemical secret, including, perhaps most obviously, the Renaissance scholarly traditions of natural magic, Hermeticism and occult natural philosophy.[3] For many in Reformation central

* My thanks to Alisha Rankin, Elaine Leong and the participants of the 2008 conference on 'Secrets and Knowledge: Medicine, Science, and Commerce 1500–1800' for their constructive comments on an earlier version of this chapter.

1 I examined these kinds of alchemical secrets in my book, *Alchemy and Authority in the Holy Roman Empire* (Chicago, IL: University of Chicago Press, 2007).

2 On the commodification of early modern natural knowledge in general, see the essays in Pamela H. Smith and Paula Findlen (eds), *Merchants and Marvels: Commerce, Science and Art in Early Modern Europe* (New York, NY: Routledge, 2002).

3 See, for example, Frances A. Yates, *Giordano Bruno and the Hermetic Tradition* (Chicago, IL: University of Chicago Press, 1964), pp. 44ff. For a more recent analysis of the association of alchemy with 'the occult' in the early modern period, see William R. Newman and Anthony Grafton, 'Introduction: The Problematic Status of Astrology and Alchemy in Premodern Europe', in William R. Newman and Anthony Grafton (eds), *Secrets*

Europe, however, it was not the world of Marsilio Ficino and Giordano Bruno, but rather the culture of apocalyptic expectation associated with Paracelsus that supplied the most powerful connection between alchemy and divine wisdom.[4] My point of departure into this Reformation context for early modern alchemy is Anna Zieglerin (c. 1545–75), who made her mark at the court of Duke Julius of Braunschweig-Wolfenbüttel in the 1570s with a distinct brand of holy alchemy.[5] Alchemy's symbolic vocabulary allowed Zieglerin to make alchemical laboratory practice relevant to an intense and widespread effort to understand sacred time and when it might end. Moreover, Zieglerin viewed her alchemical secrets as revelations, privileged pieces of knowledge imparted by God that would allow her to play an active role in the events of the Last Days. As Zieglerin's case suggests, therefore, alchemy could offer practitioners a particularly powerful way to address the pressing anxieties of a world in which many believed that the End Times were at hand.

Re-enacting the Sacred Past

Zieglerin was born into a family of minor nobility in Saxony c. 1545.[6] After a privileged childhood in the orbit of the Dresden court of Elector August of Saxony and a marriage that ended when her first husband died in a riding accident, young Anna wound up in Gotha, where her brother was a courtier at the court of Duke Johann Friedrich II of Sachsen-Gotha. Desperate to find someone to take care of his unlucky sister, Zieglerin's brother and the duke forced her (much against her will) to marry a cross-eyed court jester, Heinrich Schombach. By all accounts, it was not a happy marriage.[7] Shortly thereafter, in 1566, the young Anna and

---

of Nature: Astrology and Alchemy in Early Modern Europe (Cambridge, MA: Harvard University Press, 2001), pp. 1–37. On the notion of alchemy as revelation in late antiquity, see William Eamon, *Science and the Secrets of Nature: Books of Secrets in Medieval and Early Modern Culture* (Princeton, NJ: Princeton University Press, 1994), esp. pp. 31–2.

4 This culture is explored in depth in Charles Webster, *Paracelsus: Medicine, Magic and Mission at the End of Time* (New Haven, CT and London: Yale University Press, 2008).

5 The Niedersächsisches Staatsarchiv Wolfenbüttel (hereafter NStAW) contains extensive material related to Zieglerin, including letters, her book of recipes, supply lists and interrogation records from her trial in 1574–5. The interrogation records contain testimony by Zieglerin herself, as well as her husband Heinrich Schomach and fellow alchemist Philipp Sömmering, among others. The records can be found in the NStAW 1 Alt 9, Nrs. 306–36.

6 For the basic details of Zieglerin's time at the Wolfenbüttel court, see Albert Rhamm, *Die betrüglichen Goldmacher am Hofe des Herzogs Julius von Braunschweig: Nach den Processakten* (Wolfenbüttel: Julius Zwißler, 1883). On Zieglerin's life in particular, see p. 70, fn. 18.

7 For example, see Verhör der Anna Maria Ziegler [F. Annen verehel(ich)ung mit Heinrich], 8 July, 1574, NStAW 1 Alt 9, Nr. 314, fols 2v–3r, in which Zieglerin explains the circumstances of their forced marriage and her unhappiness with the arrangement.

her new husband Heinrich met Philipp Sömmering at the ducal court in Gotha, where Sömmering was working as an alchemist.[8] When the city was besieged in 1567, Sömmering, Schombach and Zieglerin fled together.[9] In 1571, when Sömmering eventually found work again at the court of Julius of Braunschweig-Wolfenbüttel, he invited Schombach to work as his assistant and Anna Zieglerin came along as well.

Zieglerin, Schombach and Sömmering quickly settled in to their new life at the Wolfenbüttel court. Sömmering, who had signed a contract with Duke Julius that bound him to produce an alchemical tincture within a year, set to work in his laboratory and lodgings in an old apothecary, aided by several assistants, including Schombach.[10] Zieglerin, meanwhile, seems to have served the ducal family as something like a personal shopper, procuring hats, clothing and even 'Spanish boots' for the two young princes at court.[11]

From the moment the trio arrived in Wolfenbüttel, they attracted a great deal of suspicion, as rumours of moral, legal and fiscal debauchery swirled around them. Duchess Hedwig quickly targeted Zieglerin as a particularly wicked influence at court; over the next few years she repeatedly lambasted Zieglerin's moral and spiritual rectitude, enlisting various extended family members in the campaign. In response, Zieglerin wrote several letters to Duke Julius, defending herself against what she saw as the duchess's unwarranted hostility.[12] This battle of the wills would continue for Zieglerin's entire stay at the Wolfenbüttel court, ensuring that, despite Zieglerin's own noble status, she would never find a legitimate place among the other women at court. In the spring of 1573, however, Zieglerin gave up on her attempts to play the court lady, and took on a different role entirely, one

8 Verhör und Aussage Sömmerings, 9 July, 1574, NStAW 1 Alt 9, Nr. 309, fol. 20r.

9 On Duke Johann Friedrich II's court and the events that led to the siege of Gotha, see Peter Elsel Starenko, *In Luther's Wake: Duke John Frederick II of Saxony, Angelic Prophecy, and the Gotha Rebellion of 1567* (PhD Thesis, University of California, Berkeley, 2002).

10 On Sömmering, see Rhamm, *Die betrüglichen Goldmacher*, as well as Nummedal, *Alchemy*, pp. 1–4 and 30–31.

11 In an undated letter to Duke Julius, Zieglerin reported on a shopping trip to Goslar that involved buying clothes, including the 'Spanish boots'. Anna Zieglerin to Duke Julius, s.d., NStAW 1 Alt 9, Nr. 306, fol. 9r. She and Philipp were also involved in buying a magical lucky hat for Duke Julius. Verhör der Anna Maria Ziegler [Illimij. Hut, Stunde Mercurij, and Pellitius], 8 July 1574, NStAW 1 Alt 9, Nr. 314, fols 4r–5r. On this episode, see also also Rhamm, *Die betrüglichen Goldmacher*, p. 16.

12 Anna Maria Zieglerin to Duke Julius, 12 December 1571, NStAW 1 Alt 9, Nr. 306, fol. 4r–v; Anna Maria Zieglerin to Duke Julius, 7 September 1572, NStAW 1 Alt 9, Nr. 306, fols 6r–7v; correspondence between Duchess Hedwig and Katharina, Markgräfin zu Brandenburg, geb. Prinzessin zu Braunschweig und Lüneburg, April–June 1573, NStAW 1 Alt 9, Nr. 307, fols 39r–58v. For Philipp Sömmering's perception of the conflict between Zieglerin and Hedwig, see his diary from the Markgräfin Katharina's visit to Wolfenbüttel in November of 1572, NStAW 1 Alt 9, Nr. 307, fols 177r–178v.

in which she found spectacular and perhaps unexpected success. After a year and a half at Duke Julius's court, she revealed something extraordinary: she was, in fact, an alchemist.

Zieglerin divulged her secret slowly, and by March, 1573, the court was abuzz with news of her acquaintance with an alchemical adept.[13] The fullest articulation of her alchemical secrets, however, appeared in a 20-page collection of recipes, which she presented in manuscript to the duke on 1 April, 1573, involving the philosophers' stone and a golden oil she called the lion's blood.[14] The first few lines of Zieglerin's short booklet hinted that its pages contained precious secrets and she did not disappoint.[15] With this short text, Zieglerin demonstrated her alchemical expertise, describing a number of different medical and metallurgical substances, as well as the detailed laboratory operations required to create them. These recipes alone were certainly enough to pique Duke Julius's interest, and he responded with enthusiasm; his subsequent correspondence with Zieglerin no longer focused on Duchess Hedwig's ire, but on alchemy and Zieglerin's laboratory work.[16]

These recipes are a crucial source for understanding not only Anna Zieglerin's engagement with alchemy, but also the role of secrecy in her own alchemical persona. The centrepiece of Zieglerin's alchemy was the golden oil she called the lion's blood. Although it did share certain general qualities with other alchemical substances, the lion's blood is not something that I have been able to trace exactly to other alchemical texts; rather, it appears to have been Zieglerin's signature ingredient, a secret that only she could offer her patron. The bulk of her recipes using the lion's blood had to do with reproduction, that is, with replicating in the laboratory the generative processes that early modern Europeans saw throughout the terrestrial world, in metals, plants, animals and even humans. In a nutshell,

---

13 Zieglerin discussed these rumours about the adept, Count Carl von Oettingen, in a letter: Anna Maria Zieglerin to Duke Julius, 21 March 1573, NStAW 1 Alt 9, Nr. 307, fols 77r–78r.

14 The booklet itself only exists today in a copy, which includes a note at the end explaining that, 'dieß hadtt s.f.g. bekommen zu [aus]schreyben um den ersten apprillyß anno 73 zu der vestunge Wolffenbuttel ifg ist gehandigett worden dur[ch] ifg geheymsten leybe und kammer dyenner mytt nahmen heyste der rubrechte lobrinne. Und ifg haben dißer auch auß geschrieben mytt eygen handt und angefangen den 19 huiuß und den 20 gecundigett ihn eynnygem schryben [Your Grace received it to copy out on the first of April in the year (15)73 at the Wolfenbüttel castle; it was handed to Your Grace by his private bodyguard and servant with the name Rubrecht Lobrinne, and Your Grace copied it out with his own hand and in his own writing beginning on the 19th of this month and ending on the 20th]. Anna Maria Zieglerin, 'die Edele und Tewere Kunst Alchamia belangende', 1 April 1573, NStAW 1 Alt 9, Nr. 306, fol. 70r. Rather than modernizing the German in all quotes, I have retained the orthography of the manuscript sources.

15 Ibid., fol. 52r.

16 For example, Anna Maria Zieglerin to Duke Julius, September 3, 1573, NStAW 1 Alt 9, Nr. 307, fol. 72r, where she discusses her work on 'das groste steinlein' [the greatest little stone].

Zieglerin's lion's blood was a kind of bottled essence of the generative and reproductive power that, according to some alchemical theories, lay in all of nature, but was particularly potent in metals.[17] The lion's blood, for example, could be used to grow summer fruit in winter, to create gemstones, to cure infertility and to nourish growing infants. The power of Zieglerin's oil lay in its ability to stimulate metals, plants and people to proliferate and produce 'fruchtte' [fruits] (a term she used to refer to not only literal fruits like cherries or grapes, but also human embryos or early stages of the philosophers' stone).[18] The lion's blood could do its work even when nature protested, as in the case of an infertile couple. It required only heat – which could come from decomposing dung, philosophical fire or even a woman's womb – to encourage the putrefaction that was the starting point for all generation. The lion's blood, essentially, was a distillation of the reproductive power that drove human, vegetable *and* alchemical reproduction in all of nature.

In offering her patron recipes for making and using the lion's blood, Zieglerin skillfully walked the thin line between secrecy and disclosure that ran through many similar efforts to entice patrons to support alchemical projects. She divulged a great deal in her booklet, enough to gain Julius's support over the vociferous objections of his wife Hedwig. But Zieglerin did not give away all of her secrets. Some things, she noted, could not be conveyed in writing, but required demonstration in person. In order to complete one of her recipes for the philosophers' stone, for instance, it was necessary to place two small alchemically-produced stones with some quintessence in a flask, seal it firmly, and put the flask in a philosophical fire. The secret of the philosophical fire, however, would have to wait, for Zieglerin then added, 'I will not write at the moment how it must be prepared, because I must show Your Grace in person'.[19] Moreover, a careful reader of her collection of recipes would have realized that even as Zieglerin disclosed the process for making the lion's blood, she managed to hold back one crucial piece of information. The first ingredient, it turned out, was a 'tinged gold' made by adding a certain brown powder or tincture to lead.[20] Zieglerin's booklet never revealed the contents of this powder or tincture, however, ensuring that Duke Julius could not make the lion's blood (or, for that matter, carry out most of the tantalizing alchemical processes

---

17 Zieglerin's fellow alchemist Philipp Sömmering articulated this theory of the 'vermehrende Kraft' locked up in all natural objects in a letter to Duke Julius, 25 August 1573, NStAW 1 Alt 9, Nr. 306, fols 74r–75v.

18 For example, Anna Maria Zieglerin, 'die Edele und Tewere Kunst Alchamia belangende', 1 April 1573, NStAW 1 Alt 9, Nr. 306, fol. 59v: 'laß daß wercke [unverweckt], biß ahn 20 dagen stehen, ßo kombt der frucht dyßehelben' [leave the work (undisturbed) until it stands for 20 days; then its fruits will appear]. See also fol. 65r: 'Wan man auch zu winters zeyt wille gutte reyffe zeytige fruchtte …' [If in winter time one wants to have good, ripe, fruit early …].

19 'dß ßelbige ich nycht schreyben ijeze malß, dawie eß zu gerychtte muß werden den Ich muß Ifg ßelber weyßen'. Ibid., fols 58v–59r.

20 Ibid., fol. 54r.

in her booklet) without her further instruction. Zieglerin, in other words, revealed just enough to entice her patron into supporting her, but not so much that she rendered her continued presence at court irrelevant.

All of this was rather typical for alchemists and other peddlers of secrets seeking princely patronage in the Holy Roman Empire.[21] Like the rest of them, Ziegerlin approached her patron with a secret signature ingredient or technique that promised medical and even economic benefits. She revealed just enough detail to make her claims convincing and to entice her patron to invest in her processes, but not so much that she gave it all away. In the end, in fact, it was the secrets she *kept* about the lion's blood, the details she refused to disclose, that made it valuable to her, as someone seeking patronage and status at the Wolfenbüttel court.

Zieglerin's recipes also operated on another level, however, which allowed her to frame her lion's blood as a different kind of secret – one that circulated in an economy of the sacred, as well as in an economy of health and wealth. She hinted at this additional layer of meaning in her discussion of a recipe involving a little bird that ingested the lion's blood before giving its life to become the philosophers' stone. She began by signalling the Christian context for this little bird's sacrifice: 'If you also want to … draw a heavenly parallel, then take a little forest bird and put him in a little basket'. After giving the bird nothing to eat for six weeks but a few drops of the lion's blood each day in his water, the alchemist should 'roast this little bird until it becomes like brown glass', then shatter it and pulverize it with a mortar and pestle. She concluded, 'then you have the proper *lapis* for sure(;) you can tinge greatly and copiously with this (powder?)'.[22] At the end of the recipe, Zieglerin underscored this 'heavenly parallel' again, this time more explicitly: 'on the timber of the holy cross the son of God tinged all of us poor sinners with his most holy rose-coloured blood and took part (in earthly affairs), just as this small bird will take part in the earthly tincture'.[23] Here Zieglerin made explicit another level of meaning for her lion's blood; no longer 'merely' an alchemical oil, the lion's blood also recalled the blood that Christ shed for humanity, just as the little bird's death re-enacted Christ's sacrifice on the cross.[24] Although this passage was short, and Zieglerin did not go on to elaborate

---

21 Nummedal, *Alchemy*, pp. 98–109.

22 'Wylttu auch … wolle eyn hymlische vorgleychunge machen ßo nyme eyn kleynneß waltte fogelleynne ßetz eß In eyn korbchen…bratte eß biß eß zue eyne brawnen glaß wirtt… So hastu eben ßowol dehn rechtten lapiden alß festen[;] den kanstu ßo hoche und ville myt dyße [pölffer?] dÿngiren'. Anna Maria Zieglerin, 'die Edele und Tewere Kunst Alchamia belangende', 1 April 1573, NStAW 1 Alt 9, Nr. 306, fols 54v–55r.

23 'den der ßon gottes am stamme deß heyllygen creutzes unß arme ßundern alle myt ßeynen aller heyllygeste Roßen farben blutte dyngiertte und teylhafftigk gemacht alß wirdt dieß fogeleyn der yrdischer tyncktur teylhafft'. Ibid., fol. 55r-v.

24 On the emergence of 'blood piety' in fifteenth-century Germany, see, most recently, Caroline Walker Bynum, *Wonderful Blood: Theology and Practice in Late Medieval Northern Germany and Beyond* (Philadelphia, PA: University of Pennsylvania Press, 2007).

on this 'heavenly parallel', her readers would have understood it as a metaphor packed with meaning. Above all, the parallel suggested that the alchemist could re-enact the central events of Christian soteriology by manipulating natural objects in the alchemical laboratory. The value of Zieglerin's alchemical secrets, in other words, derived from their ability not just to produce precious objects, but also to commemorate the sacred past.

Zieglerin was by no means the first to make the connection between the philosophers' stone and Christ. Just as Christ had to be crucified in order to redeem Christians, the logic went, so too must the alchemist's raw materials undergo similar torments, even death and resurrection, before they could in turn 'redeem' base metals as the philosophers' stone. Likewise, just as Christ could cleanse the soul, so too could the philosophers' stone cleanse metals, absolving them of their impurities and transforming them into the noble metals, silver and gold. As both Carl Jung and, more recently, Karl Hoheisel have pointed out, this analogy appeared frequently in late medieval alchemical texts (including the early fourteenth-century *De secretis naturae* ascribed to Arnald von Villanova, the *Codicillus* ascribed to Raimund Lull, and Petrus Bonus's *Margarita pretiosa novella*). Alchemical authors continued to draw this parallel into the seventeenth century, articulating it visually as well as textually in print.[25]

Alchemical authors employed the 'Christ-*lapis* parallel', as Jung called it, in all kinds of different ways.[26] For Zieglerin, this 'heavenly parallel' was fairly simple. It allowed her to emphasize the power of the lion's blood to cleanse impurities and thereby redeem, just as Christ's blood redeemed humanity from sin. This may explain Zieglerin's mention of leprosy, which was the only medical application of the lion's blood she included in the booklet other than recipes for using it to conceive a child. 'If you want to cleanse a leper', she wrote, 'then every day for nine entire days ... give him three drops of the oil that you extracted from the tinged gold (that is, the lion's blood); on the ninth day, tap one of his veins, and the leprosy will run out of him, or rather out of the incision, like grains of sand'.[27]

25 Carl Gustav Jung, *Psychologie und Alchemie*, vol. 2 (Olten: Walter, 1984), pp. 118–29; Karl Hoheisel, 'Christus und der philosophische Stein. Alchemie als über- und nichtchristlicher Heilsweg', in Christoph Meinel (ed.), *Die Alchemie in der europäischen Kultur- und Wissenschaftsgeschichte* (Wiesbaden: Harrassowitz, 1986); Urszula Szulakowska, *The Sacrificial Body and the Day of Doom: Alchemy and Apocalyptic Discourse in the Protestant Reformation* (Leiden and Boston, MA: Brill, 2006).

26 Jung, *Psychologie*, p. 118.

27 'Erstlych wen du eynen außßetzygen wilste reynne machen, ßo gib Ihmme neunner gantzer tage und eyn jdeß tage ... 3 troppen deß olleß ßo du von dem dingertten golden getzogen[;] amme 9 tage laß Inne eyn ader schlagen, ßo loffte dehr außßatze von Imme oder auß den Aderlaßeloche mytt herrauße alß ßandte körnner'. Anna Maria Zieglerin, 'die Edele und Tewere Kunst Alchamia belangende', 1 April 1573, NStAW 1 Alt 9, Nr. 306, fol. 63v. The description of leprous blood as having the consistency of grains of sand was common. The fourteenth-century surgeon Henri Mondeville, for instance, wrote that practitioners could diagnose a patient by handling his or her blood: 'If, when you rub this

One could read this reference to leprosy on multiple levels. Alchemical authors sometimes referred to base metals as 'lepers' that needed to be purified, invoking the biblical story of Naaman the leper (2 Kings 5:10–14), who dips seven times in the river Jordan and is thereby healed, just as the 'leprous' metals are cleansed by mercurial waters.[28] Zieglerin may also have intended her lion's blood to cure actual human lepers, however, by releasing the disease from their bodies. Zieglerin's cure for leprosy would have gestured, too, at the alchemist's holiness, just as Christ's holiness was proven by his ability to cure lepers in the Gospels.[29] Finally, leprosy's double sense as a disease of the body and soul raises the possibility that Zieglerin may have conceived of her lion's blood as something able to redeem and 'purify' not only bodies, whether of metals or lepers, but also their souls.

By hinting at the 'heavenly parallels' between the lion's blood and the cleansing, purifying powers of Christ, Zieglerin situated the alchemical reproduction and redemption of metals within a larger Biblical framework and refigured her alchemy as God's work. The real key to her holy alchemy, however, was not the metaphors threaded throughout her recipes, as crucial as those were in communicating the power of the lion's blood, but rather the mysterious adept, named Count Carl von Oettingen, whose existence was the subject of court gossip in March of 1573, just before Zieglerin presented her recipes to Duke Julius. The figure of Count Carl made it clear that Zieglerin intended her holy alchemy not merely to *re-enact* the central events of the Christian past (Christ's sacrifice on behalf of humanity), but also to *act*, to bring about new events in that salvific drama and to move it towards its immanent conclusion.

---

blood between your fingers and the palm of your hand, you find grains like grains of millet, sand or gravel, this is a sure sign of incipient leprosy'. E. Nicaise (ed.), *Chirurgie de Maître Henri de Mondeville* (Paris: F. Alcan, 1893), p. 550, as cited in Marie-Christine Pouchelle, *The Body and Surgery in the Middle Ages* (New Brunswick, NJ: Rutgers University Press, 1990), p. 73.

[28] Emblem XIII in Michael Maier's *Atalanta fugiens*, for example, depicted this scene, and explained its alchemical parallel with the following motto: 'Aes Philosophorum hydropsicum est, & vult lavari septies in fluvio, ut Naaman leprosus in Jordane' [The ore of the philosophers is dropsical and wants to be washed seven times in the river, just as Naaman, the leper, washed in the Jordan]. I have consulted the reproduction and translation in Helena Maria Elisabeth de Jong, *Michael Maier's Atalanta Fugiens. Sources of an Alchemical Book of Emblems* (Leiden: Brill, 1969), Emblem XIII (fig. 13).

[29] Matthew 8:3, Luke 5:13, and Luke 17:13–14. Leviticus 14:49–53, which discusses how a priest should cleanse the house of a leper, suggests another possible source for Zieglerin's little bird: 'For the cleansing of the house he shall take two birds, with cedarwood and crimson yarn and hyssop, and shall slaughter one of the birds over fresh water in an earthen vessel, and shall take the cedarwood and the hyssop and the crimson yarn, along with the living bird, and dip them in the blood of the slaughtered bird and the fresh water, and sprinkle the house seven times. Thus he shall cleanse the house with the blood of the bird…'. New Revised Standard Version.

Paracelsian Prophesy

In her book of recipes and in subsequent statements to the court, Zieglerin maintained that it was this man Carl, in fact, who taught her how to make and use the lion's blood. Zieglerin maintained that when she was growing up the Count had come often to her mother's house, where he transmuted lead into gold on more than one occasion and eventually served as her alchemical tutor.[30] But where did Count Carl gain his expertise? According to Zieglerin, Carl inherited his alchemical secrets from his father, who was none other than the physician and lay theologian Paracelsus (1493–1541).[31]

Zieglerin was quite savvy to endow her Count with such a famous father, of course. Moreover, she was not simply an acquaintance of the Count's. She claimed that Count Carl had been madly in love with her for years, even turning down a proposal from Queen Elizabeth of England because, as she explained, 'the Count wanted no one but her'.[32] Zieglerin even claimed that Count Carl wanted to marry her, and she forged love letters from her mysterious suitor to prove it.[33] The Count had not merely imparted his wisdom to Zieglerin, in other words, but (according to Zieglerin) also wanted to formalize their partnership in the conventional form of a marriage.[34]

We might be tempted to interpret this story merely as the romantic fantasy of a desperate woman in an unhappy marriage. But Zieglerin surrounded this powerful narrative with another equally compelling story that suggests she had something more significant in mind. The way Anna Zieglerin told it, both she and Carl were

30 Verhör der Anna Maria Ziegler, 8 July 1574, NStAW 1 Alt 9, Nr. 314, fol. 16v.

31 For a recent study of Paracelsus's life and ideas, see Webster, *Paracelsus*.

32 '…der graf niemand dan sie … wolt'. Verhör der Anna Maria Ziegler, 8 July 1574, NStAW 1 Alt 9, Nr. 314, fol. 14. See also Gütliche Aussage des Philipp Sömmering, NStAW 1 Alt 9, Nr. 308, fol. 63: 'Ehr [Carl] wolt aber keine ander haben, den die fraw anna' [he wanted no one but Frau Anna].

33 Both Sömmering and Zieglerin said that the Count even offered Schombach gold in exchange for Zieglerin. See Verhör der Anna Maria Ziegler, 8 July 1574, NStAW 1 Alt 9, Nr. 314, fol. 14, and Gütliche Aussage des Philipp Sömmering [Der Graf mit Schombach umbzusetzen], 9 [July] 157[4], NStAW 1 Alt 9, Nr. 308, fol. 64v. Unfortunately, Zieglerin later claimed to have destroyed the love letters. Her husband, Heinrich, said, 'Er wisse nit anders den das es … briffe gewesen und ungeferlich inhalts Liebe [he knew nothing other than that they were letters and more or less about love]'. Aussagen Heinrich Schombach gegen Anna, 6 July 1574, NStAW 1 Alt 9, Nr. 314, fol. 79v.

34 Anna's repeated emphasis on this point suggests the importance of the household for early modern female practitioners of alchemy. On the relationship between the household and the practice of natural philosophy, see especially Deborah E. Harkness, 'Managing an Experimental Household: The Dees of Mortlake and the Practice of Natural Philosophy', *Isis* 88/2 (1997): pp. 247–62. I examined the relationship between Zieglerin's imagined domestic arrangement with Count Carl and her claims to authority in my 'Alchemical Reproduction and the Career of Anna Maria Zielgerin', *Ambix* 48/2 (2001): pp. 56–68.

born under circumstances that pointed to their extraordinary destinies. When she was born prematurely in Saxony c. 1550, she claimed, Elector August had her wrapped in a skin 'from a woman's body' that had been rubbed with a special balsam or tincture. Baby Anna spent another 12 weeks incubating in this skin until she was fully matured.[35] Zieglerin apparently felt that her unusual birth indicated a certain spiritual virtue. As her husband later noted, Zieglerin claimed that because of this birth, 'she did not have the flux (in other words, she did not menstruate) and was more pious than others and that she wanted to be like the angels'.[36] Zieglerin's fellow alchemist Philipp Sömmering confirmed this in his own comments on her *Reinigkeit* (purity). He said that Anna frequently compared herself to the Virgin Mary, and that he too saw her as a new Mary.[37] Count Carl's birth, meanwhile, was attended by a series of portents and wonders, including a white eagle that fell out of the sky in front of Carl's mother, then rose again back up to the heavens.[38]

---

[35] 'aus eines weibs leib'. Gütliche Aussage des Philipp Sömmering, 9 July 1574, NStAW 1 Alt 9, Nr. 308, fol. 60. In trial testimony, Anna's husband Heinrich described these events in more detail. See Aussagen Heinrich Schombach gegen Anna, 5 July 1574, NStAW 1 Alt 9, Nr. 314, fols 68v–69r.

[36] 'sie den fluß nit hette und anderer mehr tugender und das sie den Engeln gleich sein sole'. Aussagen Heinrich Schombach gegen Anna, 5 July 1574, NStAW Nr. 314, fols 68v–69v. For Philipp's testimony on this point, see Gütliche Aussage des Philipp Sömmering [educatio Anna], NStAW 1 Alt 9, Nr. 308, fols 19v–20. Anna herself does not speak of this in her testimony. On this association of unusual births with special spiritual characteristics, see, for example, Carlo Ginzburg, *The Night Battles: Witchcraft & Agrarian Cults in the Sixteenth & Seventeenth Centuries*, trans. John & Anne Tedeschi (Baltimore, MO: The Johns Hopkins University Press, 1983).

[37] Gütliche Aussage des Philipp Sömmering, 9 July 1574, NStAW 1 Alt 9, Nr. 308, fol. 60. According to Philipp's same testimony, Anna attributed great significance to the fact that she did not menstruate. He claimed: 'Fraw Anna habe sich fur rein ausgeben, das sie keine menstrua hette, und wen man sich zu ihr hielte, so konten die, so von andern Menstruasisch weibern infictirt, von Ihr gereiniget werden' [Anna presented herself as pure, that she had no *menstrua*, and … those who were infected by other menstrual women would be cleansed by her]. Ibid., fol. 51.

[38] In his testimony, Philipp Sömmering recounted Zieglerin's version of events: 'Alß die Grafin diese vermeinten grafen Mutter vom Theophrasto geschwangert worden, were die nach des Kaysers hof gezog und gesagt, dass ein solch kindt von der gräfin wurde geboren werden, welcher Theophrastum Hermetem und alle Philosophos ubertreffen wurde' [And … when the Countess [von Oettingen], the mother of the supposed Count, was impregnated by Theophrastus [Paracelsus], she went to the Emperor's court and said that a child would be born from the Countess who would exceed Theophrastus, Hermes and all philosophers]. This prediction of greatness was confirmed by a sign at the time of Carl's birth. Sömmering recounted: 'Und … were ein weissen adler das aus von dem himmel gefallen. Und wie sie darnach herauß geritten, hetten sie den adler an der erden liegen sehen, und der hett sich wieder auf den himmel erhoben' [A white eagle fell out of the heavens. And when she (the Countess) went out riding, she saw the eagle lying there on the ground, and then it raised itself again up to the heavens]. Ibid., fol. 62.

Anna Zieglerin and Count Carl thus shared the fact that unusual circumstances accompanied their births. Even more importantly, however, they were also linked by a prophecy written down by Carl's putative father Paracelsus. As Zieglerin's fellow alchemist Philipp Sömmering explained: 'When the Count grew up and got his father Theophrastus's books, he found there that his father was engaged with the highest art, and that a maiden would be born who would be free of the monthly flux, and with the (maiden) the Count was to have children who would live until the final days'.[39] According to her beleaguered husband Heinrich, Anna Zieglerin believed that she was this maiden (since she claimed that she did not menstruate and was thus unusually pure, like a new Virgin Mary), destined to join Count Carl in fulfilling this Paracelsian prophecy.

This was not just a father's idle wish for innumerable grandchildren; Paracelsus's grandchildren, produced by Anna's union with Carl, would be special. Zieglerin's husband Heinrich reported her claims about these children as follows:

> His wife [Anna] also professed that the Count could create a child with her every month, and if a child lay in its mother's body for six weeks or less, it was supposed to be finished and would be raised further by the Count with the tincture; and from such children would arise a new world. The children would grow up faster than others, the females were not to be burdened with the flux … and if Adam's fall had not happened, they could well have lived forever, but now they will become as old as Methusala and other patriarchs.[40]

Zieglerin herself was a bit more modest, claiming that the Count said that their children would be born in just over six months, rather than six weeks, as Heinrich claimed. But she confirmed that the babies would be nourished after their premature birth with the lion's blood, would never become ill, and would die only when God willed it. With these children, she pronounced, she and the Count would 'prepare a new world'.[41]

---

[39] 'Alß der graff nun erwachsen und seines vaters Theophrastij bucher bekommen, hett ehr darin gefunden, dass sein vater mit höchster kunst darnach getrachtet, das ein Megdlein mucht geborn werden, die des Monatlichen flußes entf[reiet], und mit der … der graff kunder zeugen wollen, die biß Immer Jungsten tage leben sollten'. Ibid., fols 61–3.

[40] 'Seine Fraw hatte auch furgeben, das der Graffe alle monat kund mit Ir zeugen konte und wen ein kundt 6 wochen od weniger in Mutter leibe gelegen, solte volkomlich sein und durch die tinctur vom Graff weiter erzogen werden und von solcher Kundern eine newe welt werden, die kinder sollen auch eher erwachssen sein den ander, die feminae solten mit dem fluß nicht gehafft sein, … und wen sich der fall mit adam nich hette zugetragen, konten sie woll ewig leben, aber demnach wurden sie so alt werden wie Methusalem und ander patriarchen'. Aussagen Heinrich Schombach gegen Anna [Newe Welt durch Fraw Annen Kund anzurichten] NStAW 1 Alt 9, Nr. 314, fols 73v–74v.

[41] '…ein newe Welt anrichten'. Verhör der Anna Maria Ziegler, 8 July 1574, NStAW 1 Alt 9, Nr. 314, fols 14v–15r.

Paracelsus had written all of this down in a book (described by Zieglerin as both a *Kunstbuch* and a *Cabala*), which Count Carl had received from his mother and then passed on to Zieglerin.[42] Sömmering had evidently seen the book as well: ‘Frau Anna also showed him a book of the angels, how the angels in Paradise worshipped God, and that in God’s council it was decided that He wanted to let mankind fall, and then Anna and the Count would create the world again and help mankind’.[43] According to this astonishing Paracelsian prophecy, therefore, Anna Zieglerin and Count Carl would use the lion’s blood to usher in nothing less than the final act in the drama of Christian history. Just as the Virgin Mary’s body had once served as the vessel for the incarnation of Christ, with the help of the lion’s blood, this ‘new Virgin Mary’, Anna Zieglerin, would use her body to produce innumerable children who would repopulate the world and bring about the Last Days. Anna Zieglerin’s real secret, therefore, was not just a recipe for the philosophers’ stone, but rather an apocalyptic Paracelsian prophecy: her alchemy and her body would usher in the end of the world.

Alchemical Interventions at the End of Time

Today Zieglerin’s prophecy might seem unusual, even fantastic, but in its context it was not outrageous. In linking her knowledge of nature to the Last Days, in fact, Zieglerin joined many others in the sixteenth century. Robin Barnes, Charles Webster and others have noted that in the wake of the Protestant Reformation in the Holy Roman Empire, many were convinced that the end of the world would soon be at hand.[44] Anna Zieglerin’s life and ideas were deeply rooted in this

---

42 ‘Sie habe ein ganz kunstbuch gehabt, das habe Theophrastus <gemacht und> geschrieben … sie habe es vom Graf bekommen, der Graf von seiner Mutter, weiß nicht wieviel hundert stucken darin gewesen. Man [nennt] es die Cabala. Es sei auch von der alchimisterey und von constelliren darin gewesen’, as well as recipes for making clothing and curing illnesses that might kill sheep and horses. [She had an entire Kunstbuch that Theophrastus created and wrote… She got it from the Count, the Count from his mother; [she] doesn’t know how many hundreds of pieces [i.e., recipes] were in it. One calls it the cabala. It contained alchemy and astrology …] Zieglerin eventually burned this too, along with the purported love letters from Count Carl. Ibid., fol. 4v.

43 ‘‘Fraw Anna habe Ihme auch ein Buch von den Engeln gezeigt, Wie die Engele [in] Paradeiß Godt angebetten, und das in rhadt Gottes geschlossen, dass Godt denn Menschen fallen lassen wollte, und also durch den graffen und fraw annen die welt wieder anrichten, und den Menschen helffen’. Gütliche Aussage des Philipp Sömmering, 9 July 1574, NStAW 1 Alt 9, Nr. 308, fol. 42.

44 Robin Bruce Barnes, *Prophecy and Gnosis: Apocalypticism in the Wake of the German Reformation* (Stanford, CA: Stanford University Press, 1988); Deborah E. Harkness, *John Dee’s Conversations with Angels: Cabala, Alchemy, and the End of Nature* (Cambridge and New York, NY: Cambridge University Press, 1999), pp. 148–9; Webster, *Paracelsus*, pp. 8–9, 210–43.

climate of heightened expectation of the arrival of the Last Days. Her sense of the coming apocalypse likely came in part from her years in Saxony, a Lutheran stronghold that was associated with particularly fervent efforts to decipher God's plan in the Book of Nature and the Bible. This heightened spiritual awareness was no doubt strengthened during her years in Gotha, where Duke Johann Friedrich II and his close advisor Wilhelm von Grumbach communicated with divine spirits for over four years (from 1562–7), during which time they received over 5,000 'angelic pronouncements' on spiritual, economic and political matters from an illiterate peasant boy prophet named Hansie Henkel.[45] Although in Wolfenbüttel Zieglerin never spoke of Hansie's prophesies, her sensitivity to this culture of 'looking forward and waiting', as Barnes described it, certainly would have been informed by these experiences in Gotha.[46]

In this heightened state of anticipation, Barnes has argued, early modern scholars turned increasingly to astrology, biblical chronology and occult natural philosophy in an effort to decipher God's plan by reading divine signs in both scripture and nature.[47] These intensified efforts to glean new knowledge were complimented by a belief that before the Last Days, the perfect knowledge that had been lost with Adam and Eve's expulsion from the Garden of Eden would be restored to humanity. According to Christian and Jewish tradition, the Old-Testament prophet Elias (or Helias or Elijah) would return as the bearer of this lost Adamic knowledge to signal the coming end. Paracelsus, who was among those who predicted that the Last Days were at hand, characteristically put his own twist on this idea of the return of Elias. Whereas Elias was traditionally conceived of as a bearer of perfect knowledge of scripture, Paracelsus believed that he would bring with him perfect knowledge of nature and the arts, including, possibly, the secret of alchemical transmutation. As Paracelsus explained in his book *Von den natürlichen Dingen*, for example, nature's most precious secrets were yet to be revealed:

> Many arts are withheld from us because we have not ingratiated ourselves to God so that He would make them manifest to us. To make iron into copper is not as much as to make it into gold. Hence what is less God has allowed to emerge. What is more is still hidden up to the time of the arts of Helias when he will come. For the arts have Helias in the same way as other fields have theirs.[48]

---

45 On Duke Johann Friedrich II, Wilhelm von Grumbach, and the boy prophet Hansie Hankel, see Starenko.

46 Barnes, *Prophecy*, p. 261.

47 Ibid.

48 As cited and translated in Walter Pagel, 'The Paracelsian Elias Artista and the Alchemical Tradition', in Rosemarie Dilg-Frank (ed.), *Kreatur and Kosmos: Internationale Beiträge zur Paracelsusforschung* (Stuttgart: G. Fischer Verlag, 1981), p. 7. See also William Newman's discussion of the alchemical prophetic tradition in his *Gehennical Fire:*

In this way, the Paracelsian figure of *Elias artista* integrated alchemical transmutation into an older prophetic tradition, framing the very appearance of the philosophers' stone as a sign that the Last Days were at hand.[49]

As Paracelsus's followers anxiously looked forward to Elias's arrival and to the alchemical secrets he would reveal, Anna Zieglerin came to believe that she had a special role to play in the end times. Although she never mentioned any specific Paracelsian texts by name, she had clearly learned about Paracelsus and his ideas somewhere, most likely by word of mouth in the years she spent in Dresden and Gotha. Moreover, in her years at these courts where apocalyptic expectation and the pursuit of alchemy were tightly interwoven, she certainly would have become familiar with the prophesies around Elias. It may even be the case that she came to envision the figure of Count Carl as *Elias artista*. Like *Elias artista*, Count Carl would reveal his alchemical skills to Anna Zieglerin (and she in turn, would reveal them to her patron Duke Julius and the world). But Zieglerin also added another element to the Paracelsian prophetic tradition by creating an additional role for herself as Count Carl's partner. Carl needed Zieglerin to fulfill the promise of his lion's blood, for it was supposed to do more than transmute metals; the lion's blood was also to prepare a new world by populating it with alchemically produced children. Anna Zieglerin's own womb was central to this project. She would play the new Virgin Mary to Count Carl's Elias, claiming her special status as a woman of unusual purity, as she put it, to 'create the world again and help mankind' through her alchemical reproduction.

Zieglerin, therefore, offers a way of thinking about alchemical secrets not just as craft secrets or secrets of state (although they were often these as well), but also as divine secrets. Zieglerin's holy alchemy worked on many different levels. By drawing 'heavenly parallels' between Christ's blood and the lion's blood, Christ's sacrifice and the sacrifice of the little bird whose death was part of the production of the philosophers' stone, Zieglerin's alchemy re-enacted the pivotal event in Christian history: Christ's sacrifice and the redemption of humanity. The reproductive potential of the lion's blood, meanwhile, promised to re-enact not

---

The Lives of George Starkey, an American Alchemist in the Scientific Revolution, repr. ed. (Chicago, IL: University of Chicago Press, 2003), pp. 3–13.

49 According to Paracelsus, 'It is certain to be true that much lies in this earth of which I know nothing; of which others are also unaware. Then of this I am certain, that God will reveal much that is strange, much that has not yet been uncovered and revealed to us, things of which we are all ignorant. It is also true that nothing is hidden in this world which will not be revealed. For this reason, someone will likely come after me (whose deeds are yet unborn) and reveal these things'. As Paracelsus wrote elsewhere, 'God allows only the mundane to be imparted to us, the magnificent is obscured until the time of the coming of *Elias artista*. For the arts have just as much their Elias as religion also has'. As cited and translated in Herbert Breger, '*Elias Artista* – a Precursor of the Messiah in Natural Science', in Everett Mendelsohn and Helga Novotny (eds), *Nineteen Eighty-Four: Science between Utopia and Dystopia* (Dordrecht and Boston, MA: D. Reidell Publishing Company, 1984), p. 54.

Christ's death, but his miraculous birth of the Virgin. This re-enactment had a twist, however. This time, Mary would be replaced by Anna Zieglerin, who would give birth not to a single saviour, but a new flock of pure infants who would bring about a renovation of the world in its final moments. As befit a medicine that could simultaneously recall events of the Christian past and bring about events of the Christian future, the lion's blood itself was a *donum dei*, a precious gift from God that could heal a fallen world. Finally, Zieglerin's lion's blood was also a portent, its very appearance at the Wolfenbüttel court a fulfilment of the prophesy of Elias's return and harbinger of the coming End Times. In this sense, Zieglerin's golden oil functioned as a bridge between heaven and earth, simultaneously a sign that pointed to sacred events in the realm of the divine and a concrete object that could bring about very real benefits on earth.

Conclusion

One can well imagine the exhilaration Anna Zieglerin must have felt as she revealed these secrets in Wolfenbüttel. For Duke Julius, Zieglerin's primary audience, the practical appeal of the lion's blood is perhaps obvious. Surely Julius would have reaped direct economic rewards from the successful completion of her recipes; more generally, in an environment where so many other princes were also pursuing the alchemical production of precious medicines and metals, Julius would have gained prestige, and perhaps also political advantage, by hosting an alchemical success story at his court. Indeed, the book of recipes for the lion's blood was initially a huge success, vaulting Zieglerin to centre stage at Duke Julius's court and enabling her work in a laboratory of her own. Moreover, Julius was captivated by the idea of Count Carl, who still had not appeared in person at the court, but who had exchanged letters with Zieglerin and Duke Julius and promised to appear soon. As a sign of his appreciation, Duke Julius named Carl godfather to his son Joachim Karl, who was about a year old when Zieglerin presented her recipes.[50] Julius's embrace of Count Carl suggests that Zieglerin's holy alchemy gained some traction at the Wolfenbüttel court as well; Duke Julius appears to have been at least open to the possibility of engaging the far more cosmic events Zieglerin's alchemical secrets were intended to initiate.[51]

---

[50] Joachim Karl was born on 29 April, 1572.

[51] The way in which Zieglerin's holy alchemy fitted into the religious program of Duke Julius's court, particularly his introduction of the Lutheran faith in his territory, is a more complicated story than space here allows me to tell. For an overview of Julius's relationship to the Reformation, see Christa Graefe, 'Herzog Julius zu Braunschweig-Lüneburg–ein norddeutscher protestantischer Landesherr des 16. Jahrhunderts', in Cristia Graefe (ed.), *Staatsklugheit und Frömmigkeit: Herzog Julius zu Branschweig-Lüneburg, ein norddeutscher Landesherrr des 16. Jahrhunderts* (Weinheim: VCH Verlagsgesellschaft, Acta Humaniora, 1989), pp. 13–16.

Nevertheless, Zieglerin's prominence at Duke Julius' court was short-lived. In the summer of 1574, ducal officials arrested and interrogated her, along with her husband Schombach and several of their fellow alchemists and courtiers, for charges including murder, fraud and theft. Anna Zieglerin's interrogation, in particular, focused on a number of different charges: her practice of alchemy, medicine and magic; her relationship with Count Carl and a healer named Mutter Eyle (both of whom, the court claimed, did not actually exist); Zieglerin's own alleged adultery with multiple men; and her use of love magic to interfere with the Duke's marriage to Duchess Hedwig. Despite initial protestations of innocence, Zieglerin eventually confessed to everything (though her confession was hardly convincing: a record of her final statement has her saying, 'she confessed to, but didn't do, everything. She said such things out of fear'.[52] In February 1575, three and a half years after she had first arrived in Wolfenbüttel, Anna Zieglerin and seven others were put to death in particularly grisly executions. Zieglerin's skin was pinched with burning tongs, and she was burned to death in an iron chair, while her colleagues were drawn and quartered and broken on the wheel.[53]

Obviously, then, Anna Zieglerin's alchemical revelations had tragic results. What exactly led to the rapid reversal of her position, however, is less clear. It is tempting to read Zieglerin's fate as evidence that the Wolfenbüttel court rejected outright her most ambitious and dangerous claims: that she was a new Virgin Mary, that she possessed the secret of a powerful medicine, the lion's blood, and that she herself was the fulfilment of a Paracelsian prophecy. Perhaps the court could have accepted her claims about the lion's blood's ability to promote health and transmute metals, but the business about the Last Days was going too far. Put differently, perhaps some kinds of secrets – particularly those that might offer profits and health – made sense in the context of Julius's court, while others seemed out of place among the profit-driven initiatives of this entrepreneurial prince.[54]

On the other hand, the fact that the court circled back again and again to the charges of adultery and sorcery (particularly love magic) during Zieglerin's interrogation suggests that the ducal officials found her bid to be a holy alchemist worth taking quite seriously, for this was what set her apart from the dozens of other alchemists who petitioned Duke Julius and other princes. During the trial, in fact, each of her claims was carefully inverted, neutralizing her potential as holy alchemist. If there was no Count Carl, then he could not be the father of her

---

52 'sie hettes alles bekandt aber nit alles gethan solches were von ir aus furcht geredet worden'. Verhör der Anna Maria Ziegler, 16 November 1574, NStAW 1 Alt 9, Nr. 314, fol. 35r.

53 For a description of the arrest, trial and executions, see Rhamm, *Die betrüglichen Goldmacher*, pp. 45–58.

54 On Julius as an entrepreneurial prince, see Hans-Joachim Kraschewski, *Wirtschaftspolitik im deutschen Territorialstaat des 16. Jahrhunderts: Herzog Julius von Braunschweig-Wolfenbüttel* (Cologne and Vienna: Böhlau Verlag, 1978) and Nummedal, *Alchemy*, pp. 80–81.

infants. If Zieglerin's womb was not pure, as she repeatedly insisted, but rather, as the court claimed, polluted by her adulterous liaisons, then she could hardly claim to be a 'new Virgin Mary'. Finally, if she practiced demonic love magic, then she most certainly was not the holy alchemist she claimed to be. Ultimately, then, the court extracted Zieglerin's confession that her claims were a disappointing fabrication, leaving her with nothing but a collection of unproven recipes that still she had not managed to bring to completion.

This deliberate dismantling of Zieglerin's carefully constructed holy alchemy suggests that the court found her claims too potent to leave unaddressed. Had Zieglerin's claims been an obviously absurd fabrication, the court easily might have ignored them, safe in the belief that no one would believe the ludicrous fantasies of a madwoman. Julius and his officials, however, knew firsthand how compelling Zieglerin's holy alchemy could be, and their attention to it in the trial suggests that they wanted to make sure that her alchemical revelations would never be taken seriously again.

The lesson of Zieglerin's story, therefore, is something of a paradox. She failed spectacularly in her attempt to be a holy alchemist, yet the very fact that she was taken so seriously, even in her failure, shows how powerful the framing of alchemy as a divine secret could be. Something like the lion's blood could evoke craft secrets, secrets of state, secrets of nature and divine revelation all at once. And it was precisely alchemy's ability to participate simultaneously in Europe's sacred, commercial and intellectual economies of secrets that ensured that it would retain extraordinary value throughout the early modern period.

Chapter 7
Face Waters, Oils, Love Magic and Poison: Making and Selling Secrets in Early Modern Rome

Tessa Storey

In July 1613, a 'dry dark and ugly' middle-aged widow appeared before the principal criminal court in Rome, that of the *Governatore*.[1] Known as 'Maddalena the weaver', she lived with her daughter in Borgo Sant'Angelo, and although she did weave, in practice she made her living out of what Olwen Hufton famously described as 'an economy of makeshifts'.[2] As Maddalena herself explained, she ran 'a little hostellery, there over the street from my house, and I have two looms, and I have five small beds', which she presumably rented out along with the bed linens, a common source of earnings for women at this time.[3] Several months previously a young prostitute named Caterina Spirilla had rented two upper rooms from her, which may help to explain Maddalena's local reputation as a procuress, in itself another possible source of earnings.[4] In addition, Maddalena explained, 'I make waters for washing women's faces, and I also make oils from herbs of

[1] The Governor of the city of Rome was a papal appointee with jurisdiction over civil and criminal cases, a police force and two prisons. See Tessa Storey, *Carnal Commerce in Counter-Reformation Rome* (Cambridge: Cambridge University Press, 2008), pp. 8–9 and Thomas and Elizabeth Cohen, *Words and Deeds in Renaissance Rome: Trials before the Papal Magistrates* (Toronto: University of Toronto Press, 1993), pp. 5–20. The trial studied here is from the Archivio di Stato di Roma (hereafter ASR), Tribunale Criminale del Governatore (hereafter TCG), Processi, vol. 114, sec. XVII, 25 July 1613, fols c146r–347r. This trial has two sets of page numbers, one evidently imposed later, and I use this later series. All translations are mine. This evidence: ASR, TCG, Processi, vol. 114, sec. XVII, 25 July 1613, fol. c148v.

[2] Olwen Hufton, *The Poor of Eighteenth-Century France, 1750–1789* (Oxford: Oxford University Press, 1974), pp. 69 and 107.

[3] See Renata Ago, 'Di Cosa Si Può Fare Commercio: Mercato e Norme Sociali nella Roma Barocca', *Quaderni storici*, 91/1 (1996): pp. 13–33 (pp. 121–2). On women's work in seventeenth-century Rome, see also, Storey, *Carnal Commerce,* pp. 127–32.

[4] Caterina's testimony, ASR, TCG, Processi, vol. 114, sec. XVII, 29 July 1613, fol. c228r, and GioBattista's testimony, ibid., 1 August 1613, fol. c251r.

all kinds, and I can tell the difference between one herb and another'.[5] With this last statement, Maddalena was being economical with the truth, for she was also famed throughout the city for her knowledge of magic, particularly love magic, enchantments and spells. On suspicion that she had been involved in a scheme to make a rare form of poison, she was arrested and tried by the *Governatore*, along with a goldsmith named GioBattista Cristiano, her associate and presumed accomplice.[6]

The trial hearings took place over nine months, from July 1613 to March 1614, and the transcription runs to over 200 pages. A significant part of the evidence consisted of a lengthy report by a police spy who had gained access to her home and her confidence, whilst additional crucial evidence was given by Caterina Spirilla, the former lodger.[7] The testimony given in criminal records of this kind was taken down verbatim, and it often reveals many 'incidental' details about the daily life of those involved. Despite the difficulties associated with their use, they provide the historian with a rich resource of material with which to investigate the rather more mundane aspects of daily life.[8] My interest lies not in the accusations of love magic and diabolic pacts, nor in unravelling the murky truth of the events behind the arrest, but in Maddalena's activities as an unlicensed maker and vendor of medicinal and cosmetic preparations.

There is of course an intimate link between Maddalena's legitimate and illegitimate activities. The making of love potions, poisons and medicinal waters were fundamentally allied and overlapping spheres of interest which drew on the same set of skills and knowledge base. All her preparations drew on identical utensils and employed the same methods. The difference between licit and illicit lay in the ritualized use of words, gestures or objects used to confer power upon

---

5 'Fo dill'acque da lavar' il viso per li donne, et anco fo dill'olij di hirbe di tutti sorti, ma posso diffirentiare un'hirba dall'altra'. Maddalena's testimony, ASR, TCG, Processi, vol. 114, sec. XVII, 29 July 1613, fol. c225r.

6 According to GioBattista, he had met Maddalena two years previously outside her house, where they had fallen into conversation, and he had begun to frequent her house. Caterina claimed they used to meet on a daily basis for about an hour and a half. Given that GioBattista refers to several other women as his lovers, one infers that this was a kind of 'business' relationship. GioBattista's testimony, ASR, TCG, Processi, vol. 114, sec. XVII, fols 249v–250v.

7 Spirilla had met GioBattista at Maddalena's house and subsequently become his regular lover or '*amica*'. The spy's testimony is ibid., fols c147r–192r and Caterina's testimony is at ibid., fols c228r–232v.

8 The value and perils of using criminal records have been extensively rehearsed. On this, and on these sources in particular, see Elizabeth S. Cohen, 'Miscarriages of Apothecary Justice: Un-Separate Spaces of Work and Family in Early Modern Rome', in Sandra Cavallo and David Gentilcore (eds), *Spaces Objects and Identities in Early Modern Italian Medicine* (Oxford: Blackwell, 2008), pp. 8–32 (pp. 10–13) and Storey, *Carnal Commerce*, pp. 9–11.

magic potions, and in the actual ingredients used.[9] Poisonous herbs, chemicals, or harmful doses of otherwise 'safe' ingredients were what differentiated poisons from remedies, hence the determined attempts by the court to identify the poisonous herb used. In both cases, a knowledge of written or orally transmitted recipes or 'secrets' lay behind the ability to make these 'compositions', as the court described them. Maddalena was reputedly familiar with written secrets, whilst her associate, GioBattista, owned many texts containing secrets. This evidence – however slim – is rare and exciting proof of the wider circulation and use of secrets amongst lower-class social groups.

This case study explores a number of issues related to the making of secrets and remedies in early modern Rome, which I contextualize within a broader discussion of the evolution and function of Italian books of secrets. The chapter has three main aims. Firstly I argue that Italian recipe collections, known as *Libri di secreti* (Books of secrets) or *Libri di ricette* (Recipe books) cannot be considered as having emerged from, or as pertaining to, a female domain, in quite the same way as has been demonstrated for England and Spain. From the evidence so far, Italian 'secrets' collections appear to have emerged from the male-dominated arena of crafts production and medicine, only becoming more oriented towards female users in the mid-sixteenth century.

Secondly I argue that in contrast to scholarship for England, so far we lack the evidence to link the production of remedies with the average urban Italian household. Although some remedies were undoubtedly made by householders for personal consumption, the majority of households did not have the equipment or space to make large-scale medicine production feasible. I suggest that many householders would have obtained their remedies, waters, oils and so on from small-scale vendors, like Maddalena, who had the kitchen space and equipment to make such products.

My final aim is to shift the perspective away from elite production and consumption and highlight instead the link between printed books of secrets and commercial practices. By stressing their potential as instruction manuals for producers and vendors of 'remedies' for the open market, I reinsert practices involving the lower echelons of society into the narrative. In the second half

9 On magic rituals and spells in early modern Venice, see Federico Barbierato, *Nella stanza dei circoli: Clavicula salomonis e libri di magia a Venezia nei secoli XVII e XVIII* (Milan: S. Bonnard, 2002). On love magic particularly see Maria Pia Fantini, 'La Circolazione Clandestina dell'Orazione di Santa Marta: Un Episodio Modenese', in Gabriella Zarri (ed.), *Donna, disciplina, creanza Christiana dal XV al XVII secolo: Studi e testi a stampa* (Rome: Edizioni di Storia e Letteratura, 1996), pp. 45–65. For a fuller discussion see Guido Ruggiero, *Binding Passions: Tales of Magic, Marriage and Power at the End of the Renaissance* (Oxford: Oxford University Press, 1993). For an analysis of the ways in which medicine, religion and magic were intertwined in early modern Italy see David Gentilcore, *From Bishop to Witch. The System of the Sacred in Early Modern Terra d'Otranto* (Manchester: Manchester University Press, 1992).

of the chapter I use Maddalena's case as evidence for a discussion of the practicalities and practices of making secrets for sale in a domestic setting, and conclude by speculating on the circulation, transmission and meaning of secrets in her social milieu.

Gender and the Books of Secrets

The most recent research on collections of secrets and recipes has focused on England, Spain and Germany and has documented the activities of women, particularly gentlewomen, who were engaged in the production and exchange of remedies and recipes for cosmetics, medicines and foods for the use of family and friends.[10] Much evidence – particularly epistolary exchanges – reveals the extent to which recipes were transcribed, circulated and exchanged within their extensive social networks. Scholars therefore view household recipe collections as documenting female medical knowledge and practices.[11] Presumably as a reflection of women's participation in remedy making, some of the earliest printed household manuals in English, which appeared in the late-sixteenth century, were directed specifically at a female readership.[12]

In contrast, there is little to support the view that Italian recipe collections as a genre originated in a 'domestic' context, if by this we mean they were compiled primarily for use in the home, possibly by women or for women's use. Italian books of secrets contained instructions for making a wide range of products, from medicines and cosmetics to inks, dyes and varnishes. Judging from research by John Ferguson and William Eamon, Italian books of secrets appear to have developed

[10] Sara Pennell, 'Perfecting Practice? Women, Manuscript Recipes and Knowledge in Early Modern England', in Victoria E. Burke and Jonathan Gibson (eds), *Early Modern Women's Manuscript Writing: Selected Papers from the Trinity/Trent Colloquium* (Aldershot: Ashgate, 2004), pp. 237–58; Elaine Leong and Sara Pennell, 'Recipe Collections and the Currency of Medical Knowledge in the Early Modern "Medical Marketplace"', in Mark Jenner and Patrick Wallis (eds), *Medicine and The Market in England and its Colonies, 1450–1850* (Basingstoke: Palgrave, 2007), pp. 133–52; Alisha Rankin, 'Becoming an Expert Practitioner: Court Experimentalism and the Medical Skills of Anna of Saxony (1532–1585)', *Isis*, 98/1 (2007): pp. 23–53 (p. 98); Linda Pollock, *With Faith and Physic: The Life of a Tudor Gentlewoman, Lady Grace Mildmay, 1552–1620* (London: Collins & Brown, 1993); Elaine Leong 'Making Medicines in the Early Modern Household', *Bulletin of the History of Medicine*, 82/1 (2008): pp. 125–68 and Montserrat Cabré, 'Women or Healers? Household Practices and the Categories of Health Care in late Medieval Iberia', *Bulletin of the History of Medicine*, 82 (2008): pp. 18–51.

[11] Pennell, 'Perfecting Practice?', pp. 241–2; Cabré, 'Women or Healers?', p. 39.

[12] For example Thomas Dawson, *The Good Huswifes Jewell* (London, 1585) and Gervase Markham, *The English Hus-wife* (London, 1615).

principally with male users in mind.[13] Eamon suggests that the genre evolved in the context of the processes and practices involved in urban male artisanal culture: first the manuscripts, and later printed texts put into writing the technical processes and instructions of artisans, based on the experiential (as opposed to theoretical) knowledge of the dyer, tanner, alchemist or medical practitioner.[14]

The evidence from two separate library collections of Italian manuscript receipt books supports this view. Fifty-one Italian recipe manuscripts dating from the fifteenth to the eighteenth century held in the *Biblioteca Nazionale* in Florence have been inventoried and analyzed by Gabriella Pomaro.[15] These manuscripts were compiled over the course of centuries, many written partially in Latin, and where authorship can be established, none appear to have been written or compiled by women.[16] These collections of '*secreti*' contain a very broad mix of recipe types, but particularly those for the production of inks, dyes and colours, medicine, pharmacy and alchemy. Cosmetics and perfumery also feature, though to a lesser degree. Another collection of Italian manuscript *secreti* and *ricette* (secrets and recipes) is held by the Wellcome Library. Of 52 which were written largely in Italian and not exclusively concerned with alchemy, the authors of 18 can be identified, all of whom were men. Of these, six were known medical practitioners, three were priests and a further two were almost certainly health professionals, given comments on the costs of purchasing ingredients, expected profits and receipts for payments received.[17] Like the secrets collections in Florence, they contain a random mix of medical, cosmetic, alchemical and technical receipts useful for dyes, colours, varnishes and so on.

As noted above, medicinal recipes were a common component of manuscript secrets books, and within this broad category we can include recipes for the making of scented waters, oils and perfumes. As Sandra Cavallo and Montserrat Cabré describe in detail elsewhere in this volume, such preparations could play an important role in therapeutic and preventative medical care, as, according to a

---

13 John K. Ferguson, *Bibliographical Notes on Histories of Inventions and Books of Secrets*, 2 vols. (London: Holland Press, 1959) and William Eamon, *Science and the Secrets of Nature: Books of Secrets in Medieval and Early Modern Culture* (Princeton, NJ: Princeton University Press, 1994).

14 Eamon, *Science*, pp. 30–5, 54–8 and 112–30.

15 These volumes are held in the Fondo Palatino of the Biblioteca Nazionale in Florence. The inventory and its analysis has been compiled by Gabriella Pomaro, *I ricettari del fondo palatino della Biblioteca Nazionale Centrale di Firenze. Inventario* (Milan: Bibliografica, 1991).

16 Pomara comments on this 'veneer' of Latin 'which attempts to lend literary and scientific dignity to a genre which on the other hand expresses an empirical, artisan culture', Pomara, *I Ricettari*, p. XVIII.

17 See, for example, [Panelli, Francisco, et. al.], 'Segreti di diversi specifici', 1675–1748, Wellcome Western MS 3742; 'Ricette per diverse malattie', c1775, Wellcome Western MS 4202.

humoural understanding of the body, cleansing and perfuming of the body were implicitly 'medical' activities. Nonetheless, a purely 'cosmetic' function was also recognized and prized, as integral to men's and women's self-presentation; concealing unsightly blemishes, improving skin texture or colour and dying the hair. Although the names of these recipes often indicate whether they were intended principally to promote health, or to disguise and transform appearances, it is often impossible to separate these two functions: hence I will refer to them generically as 'medicinal–cosmetics'.

Scholars have observed that from the late Middle Ages, medicinal–cosmetics remedies were often separated from other therapeutic medical remedies, and an identification was made between cosmesis and women.[18] As a result, manuscripts for medicinal–cosmetics, perfumery and women's remedies were often collected separately and were specifically dedicated or addressed to women. There is evidence of this trend occurring in Italy as elsewhere.[19] By 1525, when the first book of secrets was published (*Dificio di ricette*), there appears to have been two separate publishing strands – general 'books of secrets' and cosmetics and perfumery texts.[20] The latter were directed initially at a male readership, presumably perfumers, but by mid-century were increasingly addressed to women.[21]

---

18 Monica Green observes that cosmetics literature was feminized very early on. Monica Green, 'The Possibilities of Literacy and the Limits of Reading: Women and the Gendering of Medical Literacy', in her *Women's Health Care in the Medieval West: Texts and Contexts* (Aldershot: Ashgate, 2000), pp. 1–76 (p. 35). These often circulated under the name of Trotula. See *The Trotula: A Medieval Compendium of Women's Medicine*, trans. and ed. M. Green (Philadelphia, PA: University of Pennsylvania Press, 2001).

19 Cabré finds several such recipe compilations dedicated to women in late medieval Iberia. Cabré, 'Women or Healers?', pp. 43–4. Two examples from Italy are given in Angelo Solerti (ed.), *Due codici di segreti* (Bologna: Tip. Della Ditta Nicola Zanichelli, 1894). One is untitled, the other is a 'Trattato di cosmetica'. Olindo Guerrini studies a manuscript called 'Libro dei segreti galanti et prima il trattato per fare diverse acque perfette', in *Ricettario galante del principio del secolo XVI* (Bologna: G. Romagnoli, 1883).

20 *Opera nova intitolata Dificio di ricette* (Venice, 1525). Eamon gives this as the first edition of this text. Eamon, *Science*, p. 127.

21 Eustachio Celebrino's perfumery manual first appeared in Latin, a language few women knew, before it was translated into Italian in 1525 as *Opera nova piacevole la quale insegna di far varie compositioni odorifere per far bella ciaschuna donna* (Venice, 1525). Giovanni Ventura Roseto's text on perfumes was dedicated to 'Virtuous women who wish to amuse themselves, and members of the perfumer's guild'. *Notandissimi secreti de l'arte profumatoria; a fare ogli, acque ... e tutta l'arte intiera* (Venice, 1555), p. 1. Filareto's *Breve raccolto di bellissimi secreti trovati in servizio et ornamento delle donne* (Florence, 1573) was for the 'use and ornament' of women'. See also Isabella Cortese, *I secreti de la signora Isabella Cortese, ne 'quali si contengono cose minerali, medicinali, arteficiose, & alchimiche, & molte de l'arte profumatoria, appartenenti a ogni gran signora* (Venice, 1574).

Research into Italian secrets collections has so far tended to focus on the authors or compilers of printed volumes, and on the relationship between these secrets and the evolution of early modern scientific practices. Little is known about the intended or actual users of these collections and the uses to which they might have been put. As William Eamon points out in his chapter in this volume, there is such variety within the genre that it is hard to generalize about the readership of secrets collections, intended or otherwise. Nonetheless he has suggested various ways in which they might have been read: for pleasure, for amusement, as 'self-help' manuals, and, most pertinently for my chapter, he notes their potential use as instruction books for those making preparations commercially, a point to which I shall return later.[22]

Whether printed books of secrets were intended overall for a mixed readership is hard to say. However, if we take authorship, dedications and the presence of recipes devoted to cosmesis as an indicator of a female public, then we might deduce that women were not seen as the primary readers or users. Only one of all the Italian books of secrets is by a woman, Isabella Cortese, and was published in 1574. In her very long title she clarifies that the book contains 'mineral, medical, artificial and alchemical things and many belonging to the art of perfumery, belonging to every great lady'. This can be read as implying that only the perfumery secrets were destined for women, though the sense is not clear. The only other collection which makes explicit reference to its intended readership is that by Rossello, published in 1576, which was for 'men and women of great intelligence'.[23]

Turning to the proportion of cosmetics recipes, an analysis of the contents of eight secrets collections printed before 1620 also suggests only a minimal bias towards a female readership.[24] I find that they account for 14 per cent of recipes in

---

22 On elite experiments, see Eamon, *Science*, pp. 148, 122 and 127.

23 Cortese, *I Secreti*, title page and Timoteo Rossello, *Della Summa de' Secreti Uniuersali in Ogni Materia* (Venice, 1575–6). See also Jo Wheeler's discussion in his *Renaissance Secrets, Recipes and Formulas* (London: V&A Publishing, 2009), p. 11.

24 Subsequent analyses of recipes and recipe books are based on the Wellcome Trust and University of Leicester funded 'The Italian Books of Secrets Database' project (Tessa Storey, David Gentilcore and Sandy Pearson). The project is accessible at https://lra.le.ac.uk/handle/2381/4335. 'The Italian Books of Secrets Database' draws on 1,390 recipes pertaining to the care of the body from the sixteenth and early seventeenth-century taken from three of the most important and widely circulated sixteenth-century books of secrets and two lesser known Chapbooks. These are: *Dificio di ricette*; *Recettario novo probatissimo a molte infirmita* (Venice, 1532); Alessio Piemontese, *De' secreti del reverendo donno Alessio Piemontese, prima parte, divisa in sei libri* (Venice, 1557); *La seconda parte de i secreti di diversi eccellentissimi uomini nuovamente raccolti* (Milan, 1559); *Dei secreti di diversi eccellentissimi uomini nuovamente raccolti. parte terza* (Milan, 1559); *De secreti del R.D. Alessio Piemontese. Parti quattro* (Venice, 1674); *Tesoro di varii secreti naturali* (Venice, 1600); Francesco Scarioni da Parma, *Centuria di secreti politici, cimichi, e naturali* (Venice, 1626). The first edition of *I secreti* part IV was

the *Dificio* and between only 2 to 3 per cent of recipes in four other printed books.[25] An exception to this pattern are the first three volumes of the *Secreti* purportedly authored by Alessio Piemontese, published in the 1550s and 1560s.[26] Cosmetics account for between 17 per cent and 30 per cent of the recipes in these volumes.[27]

This is not to argue that books of secrets were only read by or used by men. Rather, judging from the existence of a separate cosmetics genre, and from the content and dedications of secrets books, I suggest there was a widespread assumption that women were more likely to concern themselves with the making of medicinal–cosmetics and with a few simple domestic and 'women's remedies' whilst men were associated with the making of medicinal remedies and general domestic 'chemicals'. Whether this was reflected in household practices clearly merits further research. However, the only Italian secrets manuscript in the Wellcome Library which was dedicated to, and presumably used by a woman, confirms the impression that women were expected to confine themselves to medicinal-cosmetics. It was dedicated to a Caterina Medici in the convent of *Le Murate* in Florence in 1598. The 300 pages of neatly copied recipes were presented to her by Don Antonio, 'Prince of Tuscany'. They had all been 'experimented by his own hand' and were 'all true' and they were exclusively recipes for scented waters, oils, cosmetics and perfumery.[28]

Kitchens, Cookery and Making Remedies

Another question is the extent to which the average Italian householder would actually have been equipped to make their own remedies. Scholars of early modern medicine have found much evidence that the majority of healthcare in medieval and early modern Europe took place within the home and that routine health and hygiene were seen as a woman's duty and past-time, whether for her

published in 1568. See John Ferguson,'The Secrets of Alexis', Proceedings of the Royal Society of Medicine, Section on the History of Medicine, 24 (1931): pp. 225–46 (p. 244). My thanks to William Eamon for this reference.

25 Although as noted above, there is much overlap in recipe functions, I describe as purely 'cosmetic' those which aim primarily to alter one's appearance, rather than treat or remove the signs of disease. These four are the *Recettario novo*; *De secreti del R.D. Alessio Piemontese, Parti Quattro*; *Tesoro di varii secreti*; and Scarioni, *Centuria di secreti*.

26 'Alessio Piemontese' is thought to have been Girolamo Ruscelli. Eamon, *Science*, pp. 139–40.

27 The fourth volume has barely any cosmetics. *Piemontese, Secreti … Parti Quattro*.

28 'Segreti di D. Antonio de Medici … dati e donati a me Caterina Medici nel convento delle Murati di Firenze, l'anno 1598', Wellcome Western MS 485. According to the title page it contains waters, oils, foundations, hair washes, scented oils, balsams, pastes, powders and pomades, and so on. I can find no reference to a Caterina Medici who was alive in 1598.

close family or servants and other dependents.[29] Within this context, a link has been made between kitchens, kitchen utensils and techniques and the preparation of medicines. Pounding, boiling, straining, the use of skillets, pans, cauldrons and sieves, and cookery in general, have all been cited as examples of skills and objects which women could easily have transferred to the making of remedies.[30] In the English case in particular, kitchens and cookery were considered to be women's territory and research has shown that it was common for the women in gentry and aristocratic families to play a role in providing basic medical care for themselves, their families or their estate workers and tenants.[31]

Scholarship on early modern Italy so far has not addressed practices of domestic remedy making and whether they shared the same cultural expectations that linked elite women to cookery and kitchens – and hence domestic remedy making. However, in a study on the material culture of the early modern European household, Raffaella Sarti argues that in Italy and France kitchens and cookery, particularly amongst the elite, were the preserve of men, not of women.[32] In didactic texts Italian gentlewomen, far from being encouraged to involve themselves in cooking and the making of remedies, were advised not to sully themselves by association with such menial tasks in the kitchen. This conviction, therefore, also may have applied to the making of remedies.[33] Further study is needed to establish the extent to which women actually obeyed these injunctions.

Lower down the social hierarchy one assumes that women were routinely involved in food preparation, and judging from different studies based on Roman inventories it was not uncommon to own a tripod which could have been hung over a fire, a few pans, a plate or two and a glass. However rooms dedicated as

[29] Cabré stresses the importance of women in attending to the role played by the non-naturals in the prevention of illness in 'Women or Healers?', pp. 28–30. See also Sandra Cavallo, 'Health Beauty and Hygiene', in Marta Ajmar-Wollheim and Flora Dennis (eds), *At Home in Renaissance Italy* (London: V&A Publications, 2006), pp. 174–87 (p. 174). Mary E. Fissell, 'Introduction: Women, Health, and Healing in Early Modern Europe', *Bulletin of the History of Medicine*, 82/1 (2008): pp. 1–17 (pp. 1 and 13); Green, 'Possibilities of Literacy', p. 46; Susan Broomhall, *Women's Medical Work in Early Modern France* (Manchester and New York: Manchester University Press, 2004), pp. 127–38 and Gianna Pomata, *Contracting a Cure: Patients, Healers, and the Law in Early Modern Bologna*, trans. G. Pomata, R. Foy and A. Taraboletti-Segre (Baltimore, MD and London: The Johns Hopkins University Press, 1994), p. 79.

[30] Leong, 'Making Medicines', p. 162. See also Cavallo, 'Health Beauty and Hygiene', p. 176.

[31] Linda Pollock, *With Faith and Physic: The Life of a Tudor Gentlewoman, Lady Grace Mildmay, 1552–1620* (London: Collins and Brown Ltd, 1993), p. 97. Leong, 'Making Medicines', p. 146 and Rankin, 'Becoming an Expert Practitioner', p. 25.

[32] Raffaella Sarti, *Vita di casa. Abitare, mangiare, vestire nell'Europa moderna* (Rome and Bari: Editori Laterza, 2000), pp. 183–90.

[33] Ibid., pp. 183–7.

kitchens and more sophisticated kitchen utensils were usually absent. Indeed, as Renata Ago writes:

> There was evidently a band of the urban classes, sufficiently well off as to own a place to live, furniture, linen and decent clothing, who nonetheless were not able to cook in their homes and even less able to consume their own food on plates, or drinks in glasses.[34]

Research also suggests that much food was bought in from hostelleries and other street vendors rather than being prepared in homes.[35] As for the containers required for the distilling and storage of remedies, this equipment appears to have been extremely uncommon in the typical household.[36]

The lack of evidence for the presence of elite Italian women in kitchens, and of alternative spaces for the preparation of medicines, as well as the absence of all but the most basic cooking equipment from lower-class homes, raises questions both about the role played by Italian women in domestic remedy making and about the extent to which Italian remedy making actually took place in the home. If elite women were dissuaded from getting their hands dirty in the kitchen, whilst the majority of middling and poorer households appear to have lacked the spaces and utensils to make anything but the most basic of meals, let alone complex herbal preparations, who, aside from apothecaries, made medicinal remedies and cosmetics?

As regards women in the elite, the connotations of making perfumes and face washes may have been more positive than for cookery. Cortese's volume, as noted earlier, specifies that the secrets of the art of perfumery are suitable 'for every great lady'. Similarly, the titles and dedications of cosmetics tracts tend to emphasize the nobility, intelligence or 'virtuous' nature of its female readership.[37] And for those who preferred not to get their hands dirty, servants

---

34 Her analysis is based on roughly 100 inventories and accounts books from well-to-do households (only two from the nobility) and 100 wills. Renata Ago, *Il gusto delle cose* (Rome: Donzelli Editore, 2006), p. 93. On the contents of the homes of prostitutes and courtesans see Storey, *Carnal Commerce*, pp. 190–92 and 203–5.

35 Sandra Cavallo, 'The Artisan's Casa', pp. 66–75 (p. 75) and Storey, *Carnal Commerce*, pp. 203–5.

36 Ago finds only one collection of domestic distilling equipment. A separate study currently underway into the material culture of domestic preventative medicine finds likewise. This is based on analysis of 92 Roman inventories from the mid-1550s to the 1690s. Marta Ajmar, Sandra Cavallo and Tessa Storey, 'Healthy Homes and Healthy Bodies in Renaissance Italy', Wellcome Trust Funded Project. Online: http://www.rhul.ac.uk/history/Research/HealthyHomes/index.html.

37 Timoteo Rossello, for example, specifies that his book is for 'men and women of great intelligence…and all virtuous people' – when virtue was virtually synonymous with nobility. See, for example, Baldassare Castiglione's *Il libro del cortegiano* (1528), Giovanni della Casa's *Galateo* (1558) and Stefano Guazzo's *Civil conversazione* (1579).

could always be assigned the manual labour involved.[38] Some cosmetics, therefore, may have been produced in elite homes, although more research is needed on this topic. Even if this were the case, it still leaves us with a puzzle over who was expected to make cosmetics and remedies for use in lower-class homes, given the obstacles to making significant or sufficient quantities of remedies at home.

Selling Medicinal-Cosmetics in Seventeenth-Century Rome

Maddalena's cosmetics business points us towards the existence of small-scale producers who could supply householders with a broad range of common medicinal–cosmetic waters and oils.[39] Recent accounts of early modern medicine have noted the gradual commercialization of medical services and have also stressed the multiplicity of health practitioners operating within the urban environment and the consequent blurring of boundaries between one trade and another, despite attempts at regulation.[40] It is in this context that we need to view the products offered by Maddalena. As a woman making products for other women, for external use only, and for which she did not appear to have claimed any therapeutic value, Maddalena presumably fell outside the remit of licensing authorities. Furthermore, her income from these very humble products was probably too insignificant to have attracted their attention.[41] Maddalena's failure to declare to the court that she had permission to sell her products suggests that she was not alone in earning extra income in this manner. The mention of her trade in a sixteenth century novella, *La lozana andaluza*, confirms the ubiquity of such activities.[42]

In this period, the retail of herbs and remedies in Rome was dominated by the apothecaries – whose college had authority over many of the smaller less powerful guilds. Apothecaries were licensed to dispense a huge range

---

38 See, for example, Rankin, 'Becoming an Expert Practitioner', p. 24.

39 On apothecaries in Rome and Venice see Cohen, 'Miscarriages' and Filippo di Vivo, 'Pharmacies as Centres of Communication in Early Modern Venice', in Cavallo and Gentilcore, *Spaces, Objects and Identities*, pp. 33–49 (on their clientele see especially pp. 41–4).

40 See the introduction to Jenner and Wallis, *Medicine and the Market*, pp. 1–23. For the Italian case see David Gentilcore, *Healers and Healing in Early Modern Italy* (Manchester: Manchester University Press, 1998).

41 I am guided here by Gianna Pomata's discussion of female healers and their relationship to the protomedicato in Bologna in Pomata, *Contracting a Cure*, pp. 7–9.

42 Francisco Delicado, *La lozana andaluza*, ed. and trans. Luisa Orioli (Milan, Adelphi, 1970). The novel was written in Rome in 1524 and published in Venice in 1528. It describes women in Rome who earned their living by going about the city 'making up brides and selling preparations of sublimate and water for the skin', p. 32.

of simples and compounds.[43] Their main rivals were the *spetiali* (grocers), the itinerant charlatans – who sold simples as well as compound remedies – and the *aromatari* (perfumers) – who sold perfumed products such as waters, essences and unguents.[44] There were also the *sempliciste* (herbalists) who were authorized to identify and supervise the gathering of herbs for sale as well as make herbal remedies. However there was a panoply of other smaller trades associated with medicinal preparations, about whom very little is known other than their names. There were for example the *cammomillari* who gathered and prepared infusions and decoctions from camomile, the *distillatori* who made oils and essences from water and the *acquaioli*, who also sold scented waters for the face as well as drinking water from specific springs.[45]

Maddalena's numerous oils and waters must have been very similar to those sold by the *distillatori* and *acquaioli* or indeed *aromatari*, which suggests that there was a healthy market for such skin care products. This is certainly supported by evidence from David Gentilcore's study of charlatan's licenses, which finds that skin remedies formed the largest single category of medicines sold during this period. This is further reinforced by the finding that they also represented the most important single remedy category amongst medicinal recipes in the books of secrets I have studied.[46]

This brings us to the role the books of secrets could have played in providing the instruction necessary for those aspiring to, or engaged in the commercial production of remedies, cosmetics and other substances.[47] Certainly this function was foreseen by the editor of Alessio's second volume of *Secreti*. In the instructions for a skin whitener we read, 'If you want it to smell good, because you are making it for a lady, add some musk or civet', whilst a recipe for making small barber's soaps adds that 'you can sell these immediately, although they will not be worth much'.[48]

---

43 David Gentilcore, '"For the Protection of Those Who Have Both Shop and Home in This City": Relations between Italian Charlatans and Apothecaries', *Pharmacy in History*, 45/3 (2003): pp. 95–107 (p. 109). On competing claims for authority between related guilds see also Cohen, 'Miscarriages'.

44 Gentilcore, 'Protection', p. 109. On the charlatans generally see David Gentilcore, *Medical Charlatanism in Early Modern Italy* (Oxford: Oxford University Press, 2006).

45 Sandro Salvi and Giorgio Roberti, *Antiche farmacie Romane: Speziali, medici guaritori fra cronache e magia* (Rome: Fratelli Polombi, 1988), pp. 20–3.

46 Skin remedies constituted between 40 per cent and 67 per cent of all licences issued in the late sixteenth and early seventeenth century. Gentilcore, *Medical Charlatanism*, p. 206. They accounted for 22 per cent of medical recipes in the *Dificio di ricette*, 59 per cent of those in the *Recettario novo* and between 20 per cent and 32 per cent in the first three volumes of Alessio Piemontese's *Secreti*. When the 1,858 medical recipes in the database are catagorized according to 64 'ailment' categories, the single most important category is 'skin complaints'.

47 Eamon, *Science*, p. 122.

48 Piemontese, *Secreti*, vol. 2, fols 46v and 53r.

Acting outside the guild structures, which would traditionally have provided training in the art of making such 'secrets', Maddalena could have relied on the oral transmission of remedies. Books of secrets may also have been a useful source of information, given the great many recipes they contained for the kinds of products she made. How likely was it that she had come into contact with written secrets, and were there any links between her products and recipes in circulation at the time?

The Circulation of Secrets

Literacy was surprisingly common amongst Italian city dwellers, and evidence from trials in Rome, Venice and Modena shows that even prostitutes were able to write letters and spells.[49] According to her neighbour, Maddalena was no exception to this. When describing her to the police spy, her neighbour said that Maddalena was sought out by men and women of the elite. Part of her reputation rested on the fact that she was familiar with 'written secrets' – suggesting that familiarity with writing conferred authority on the secret.

> Maddalena had *gran secreti scritti*, that is what she had understood and she was known as a great and *valente* woman on account of her ability to make people love one another with enchantments and *scongiuri*, and friars, priests, gentlemen and ladies thronged her house, and other people, she was held in such high esteem.[50]

Although this is not proof, it does make it seem more likely that she was literate.

As regards access to texts, scholars have stressed the huge numbers of books, pamphlets and manuscripts containing secrets circulating at this time. A copy of Alessio's *Secreti* identified in a Roman inventory from 1587 was valued at a mere 0.15 *baiocchi*, barely the price of a small flask of wine and chunk of bread in a hostellery.[51] Even if Maddalena had not seen such a volume, there were many other sources for secrets. She might have been shown or taught her secrets

---

[49] On the literacy of prostitutes see Fantini, 'Circolazione'; Fantini, 'Les mots secrets des prostituées (Modéne, 1580–1620)', *Clio. Histoire, femmes, et sociétés*, 11 (2000): pp. 21–47; Barbierato, *Nella stanza dei circoli*, esp. pp. 86–92. See also a long letter by a Sicilian prostitute in ASR, TCG, Processi, vol. 23, 1602, fols 319–40.

[50] 'Madalena haveva gran secreti scritti che cosi haveva intesa e che lei era conosciuta per una grande et valente dona per conto di far voler bene per incanti e scongiuri et che in sua casa ci barrigava frate e prete signori et signore et altre gente esendo tenuta in gran concetto'. Spy's testimony, ASR, TCG, Processi, vol. 114, sec. XVII, 8 July 1613, fol. c148v.

[51] ASR, Trenta Notai Capitolini, uff. 4, vol. 75, 20 September 1587, fols c601r–609v.

orally; she may have seen her recipes on single printed sheets, or in a manuscript collection.[52] When her accomplice GioBattista's premises were searched, about 16 texts described as 'small printed books and handwritten orations' were found, one of which contained 'certain secrets'.[53] As a goldsmith, he would have been familiar with the skills required to make up remedies and would have had access to distilling equipment, scales, furnaces and the permission to use certain chemical ingredients. Maddalena could well have learned both techniques and recipes from him or from his books, or from others like him.

Evidence for practices of making secrets is extremely rare, and in the following section I will use the testimonies of Maddalena and the spy to draw out the parallels and similarities between her recipes and practices and those found in some contemporary texts. My first aim is to highlight the extent to which her practices belonged to the broader culture of secrets making, whether she had acquired her knowledge orally, through manuscript or through print. My second aim is to draw attention to the equipment and methods used by a small-scale domestic 'maker of secrets', like Maddalena, as a way of illustrating that although it was possible to make up these recipes in a small urban home, it seems unlikely that it was common practice. Maddalena's household and circumstances, on the other hand, were well suited to this venture. Widowed, she had an income which permitted her to live in a house with several rooms rather than the usual one or two, and which must have included some kind of kitchen facilities, since she ran a hostellery. Her income had also enabled her to amass a range of utensils suitable for distilling and storing potions, which we would not find in the average Roman household, as we will see below.

Making Secrets

The kinds of products in which Maddalena specialized could be found in any contemporary book of secrets. She mentions for example making 'waters for washing women's faces'. In books of secrets, face waters were overwhelmingly scented washes, used not only to clean the face but also to whiten and improve skin texture, or to heal skin conditions. 117 examples can be identified in the eight volumes I have studied in 'The Italian Books of Secrets Database'. Oils were also common remedies – there are 64 examples in my sample. These tended to have a more medicinal function, healing sores, boils, wounds and other skin conditions, with very few intended for purely cosmetic purposes. Maddalena

---

52 On the circulation of secrets in the Italian context see Eamon, *Science*, p. 27, and Barbierato, *Nella stanza*, pp. 83–111, 183 and 237.

53 For the inventory of what the police found of interest in his apartment see ASR, TCG, Processi, vol. 114, sec. XVII, 1 August 1613, fol. 246r–c247r. This was despite having had most of his books burned by his confessor not long before, as described by Caterina: ASR, TCG, Processi, vol. 114, sec. XVII, 29 July 1613, fol. c229r-v.

made 'oils from herbs of all kinds', naming camomile, wormwood, roses and scorpions but emphasizing particularly that she made one from St John's wort (hypericum).[54] Although a common herb, of 1,390 pre-1600 printed remedies in my database only four recipes were for an oil using St John's wort as a key ingredient. Several of the details in Maddalena's description of preparing her oil resemble these four recipes, particularly the injunction in Alessio's *Secreti* for the making of a 'red oil' from hypericum: that one should pick the herb 'on St John's day (or between the twentieth and thirtieth of [June]) after dawn'.[55] Maddalena echoes this requirement when she mentions on three separate occasions that the herb is associated with St John's day morning, as when she explains that, 'This herb, which is called St John's Wort, is very long with certain yellow flowers and it is gathered the morning of St John's day because it is one which has great virtues'.[56] The police search of her home also revealed a small vial of a 'stinking' red oil. Although she may never have laid eyes on a printed book of secrets, it is not far-fetched to suggest that somewhere along the way she had come into contact with a written or oral version of Alessio's Red Oil.[57]

According to the books of secrets, there were a variety of ways to make these kinds of recipes. Some were very simple – as in steeping herbs in a liquid – and really could be made at home. However, many were very complex, particularly cosmetics recipes. They might use many ingredients, including chemicals, extremely hot ovens and multiple glass alembics for the different phases of distillation. But even to make the simplest of these, ingredients had to be purchased or gathered, space and time were required for the different phases of preparation and the ingredients and finished products needed to be stored, sometimes for long periods, before they were ready for use. All this required time, space, money and skills which, I suggest, were not available or practical on an everyday basis to most households, but to which someone like Maddalena could devote her time and energy.

Maddalena sourced her ingredients in a number of ways, ranging from picking them herself to drawing on a network of suppliers.[58] The cheapest way to acquire

---

54 She also mentions three others which I cannot identify: *madrigali, lacriminia* and *lombriento.*

55 Piemontese, *Secreti*, vol. 1, p. 25.

56 'La hirbe sopradetta, si chiama herba di San giovanni et è lunga lunga col certi fiori gialli e si coglie la matina di San Giovanni perchi ch'uno ch'ha gran virtù'. Maddalena's testimony, ASR, TCG, Processi, vol. 114, sec. XVII, fol. c216v. See also 'L'hirbe colta la matina di San Givanni' and 'Dill'hirbe dilla matina di San Giovanni'. Ibid., fol. c215v.

57 The police found in her house 'a *carafilla* of glass, which stinks strongly of … oil, coloured red' ['una carafilla di vitro che puzza assai col olio dintro color rosso'], which suggests that her oil was also the right colour. ASR, TCG, Processi, vol. 114, sec. XVII, 27 July 1613, fol. c191v.

58 This resembles the way in which the English gentlewoman, Elizabeth Freke, went about making her medicines a century later. See Leong, 'Making Medicines', p. 167.

the ingredients was obviously to gather the herbs oneself, and twice Maddalena mentions having picked the hypericum at dawn.[59] While she fails to describe where she had picked the herb, seventeenth-century Rome had plenty of *vigne* (vineyards) and *horti* (gardens) both within and outside the city walls. Maddalena's accomplice GioBattista also collected his own ingredients and mentions that he picked bracken seeds by night when staying in the country. Peasants could also supply them with ingredients. GioBattista had traced a supplier of a rare and extremely poisonous herb known as *nepello* in the mountains, and a peasant had then brought it into the city in a chest.[60] This use of an informal supplier presumably reflected the illegality of the substance, although there were probably no official vendors for certain ingredients. Toads, slugs, earwigs and so on were part of the standard *materia medica* and may have been supplied by apothecaries, but those used by Maddalena in her remedies were brought to her by a peasant woman from the country and were kept in the cellar and fed with bread.

In comparison with the average household in Rome, Maddalena had an impressive collection of simple utensils on hand, and her testimonies suggested that she had a certain expertise. In order to make a face water, for example, she took 'file di ravina' (possibly brambles):

> Which I dissolve in water … and then I put them in the bell-jar (*campana*) and distil them, and another kind with watermelon, in which I put egg white and bread crumbs from a white loaf, and I put that in the bell-jar and distil it, and another kind I make with sublimate and tincture of *spitiana* … with water … in the bell jar and I distil it.[61]

The *campana* to which Maddalena refers was the top part of an alembic or retort, commonly called for in all kinds of medical and cosmetics recipes, and the method she favoured for distilling her oils and waters was 'without a flame, in the sun'.[62] She pointed out that the sun shone very brightly in her apartments since she had a loggia. Possibly she emphasized this very simple method because the courts were more likely to suspect those who possessed stills and burners of being implicated in illicit alchemical procedures, but her testimony was corroborated when the

59 Maddalena's testimony, ASR, TCG, Processi, vol. 114, sec. XVII, 28 July 1613, fols c215v–216r.

60 Caterina's testimony, ASR, TCG, Processi, vol. 114, sec. XVII, 29 July 1613, fols c230v–231r.

61 'Chi lo fo stimpirare col'acqua … e poi li mitto nilla campana e lo distillo et un altra sorte la fo col cocommiro e dentro ci mitto chiaro d'ovo e la mollica di una pagnotta bianca e poi la mitto nilla campana e lo distillo et un'altra sorte la fo col il colimato e col tintura di spitiana, rasura di … E col acqua … Nilla campana e la distillo'. Maddalena's testimony, ASR, TCG, Processi, vol. 114, sec. XVII, 28 July 1613, fol. c216r.

62 'Che io li distillo sinza fuoco al sole'. Ibid., fol. c215v.

police found various glass vials hanging from a wall and others above the loggia, having been left to distil in the intense heat of the summer sun.[63]

Maddalena also used some rather more sophisticated equipment. According to the testimony of the spy, there was 'a glass container like a wine flask which is covered with clay outside, in the bottom of which you place two fingers of the mixture'.[64] And according to his recollection of her instructions:

> And above this vase she puts something 'like a sponge', which contained another ingredient which draws towards itself the thing in the bottom, slowly slowly, through the strength of the heat of the sun ... until the thing in the bottom is incorporated with the thing in the top of the vase, then drop by drop, through the heat of the sun, it returns down below.[65]

The very vagueness of the spy's language when he attempts to recount this conversation suggests that neither the equipment nor the processes described were familiar to him and were therefore not common domestic tools or practices.

Using the sun was fairly simple, but the books of secrets frequently specify the use of a *bagno maria*, in which the receptacle is placed over a container of boiling water until it has reduced or distilled. This was a process which could take days rather than just hours and required a continuously burning heat source. Maddalena's description of how she distilled during the winter months reveals an ingenious homemade solution to the challenge this posed: she said that in winter she used a similar vase to that used in the summer, then made a little oven out of clay, and placed her glass vase over it. She then put a lantern under the vase 'which she doesn't allow to go out day or night, checking it often to make sure it is alight, leaving it as long as is needed'.[66] This explanation suggests either that she did not have a fireplace or that a traditional domestic fireplace was not suitable for this task.

Storage of ingredients and the finished product was also a consideration since scented waters in particular were often made in bulk, left to stand for long periods and would need to be transported when sold. Scattered around the house,

---

63 Spy's testimony, ASR, TCG, Processi, vol. 114, sec. XVII, 27 July 1613, fol. c191r-v.

64 'Una cosa di vetro, come fusse una foglietta tutta increta da di fuora. Con doi dita di una cosa, che io ci metto in fondo'. Spy's testimony, ASR, TCG, Processi, vol. 114, sec. XVII, 11 July 1613, fol. c157v-158r.

65 'E sopra a detto vaso, ci mette una cosa come se fosse una sponga, che tira a poco, quella cosa che sta in fondo per forza del gran caldo, [...] dopo haver tirato a se tutto quello che sta in fondo et incorporata con quella cosa che sta sopra il vaso; a gocia a gocia per forza del gran caldo torna a basso'. Ibid.

66 'Che mai non si smorza giorno ne notte revedendo la spesso per far la stare acesa lasciandolo star cosí quel timpo che la persona vole dare'. Spy's testimony, ASR, TCG, Processi, vol. 114, sec. XVII, 11 July 1613, fol. c160v.

Maddalena had numerous examples of her work kept in a range of containers. She had more than ten *ampolette* (vials) containing her oils; she had earthenware vases containing vinegars stored under the roof; she had a vase of vinegar above a wall on her loggia and an ampoule of St John's wort oil hanging from a nail on the wall of the loggia; a crystal ampoule with a little white unguent inside also hung there. Near it was a piece of *cirobo* (?) wrapped in linen which she said she used for her leg. She had a small jar of herbs in her bedroom with some acqua rosa in it, and other *fiaschi* (flasks), big and small.

Maddalena's business was small compared to those run by licensed practitioners who were members of guilds. Yet she made numerous standard oils and waters that could have appealed to a wide spectrum of the population.[67] She had equipment that was absent from the average household, room to prepare and store her products and obvious conversance with a range of methods. Finally, thanks to her hostellery, her weaving, her procuring activities and her reputation for magic (judging from her neighbour's accounts), she was embedded in an extensive social network through which she could market her goods, which spanned the social hierarchy and extended well beyond her immediate neighbourhood.

Secrets and Secrecy

By the time Maddalena was making her waters and oils, printed recipes for such medicinal 'secrets' had been in circulation for almost a hundred years. Yet it is interesting that although the term 'secrets' is used on numerous occasions in the trial, by different speakers, in different contexts and with several meanings, there is never any reference to these medicinal–cosmetic products as 'secrets'. This brings us to the final question, of what was meant, in the streets and courtroom in early seventeenth-century Rome, by the word 'secrets'?

The first mention of 'secrets' was made by the neighbour when she declared that Maddalena knew 'great written secrets', in a context which suggests that she was referring to her reputation for efficacious love magic. Maddalena's secrets were occult knowledge, and her neighbour's emphasis on the fact that they were written seems to imply that a text conferred a certain authority on the secret and its user, or that a written secret was considered more efficacious than one which had been merely memorized – something which has been observed by scholars working on magical practices in other parts of Italy.[68]

---

67 Compare to the case of Elizabeth Freke in Leong, 'Making Medicines', pp. 162–4. Maddalena had far fewer remedies.

68 This link between writing, books and authority is also observed by Barbierato who comments that 'ownership of a magic book and recognized ability to practice it was a guarantee of a priviliged position, 'almost of power'. Barbierato, *Nella stanza dei circoli*, pp. 83–4. Fantini makes a similar observation in 'Circolazione', p. 56.

In her direct testimony to the court, Maddalena does not use the word *secreto*, certainly not in connection with her medicinal–cosmetics, which she describes only as 'oils', 'waters' and 'herbs' or as 'compositions', echoing the vocabulary used by her interrogators. However, according to the spy's account of his conversations with Maddalena, she did use the word *secreto*, but only in connection with her occult practices. Although this means we cannot be sure that she herself uttered those very words, we can assume that these were the kind of words used in that social environment and which the spy expected a woman of her kind – as a dabbler in the occult – to use. According to him she had said, 'If I want to, I can make [someone] die with a secret and a poison which I know how to make'.[69] The implication here is that the 'secret' is a recitation of magical words – a spell or an *oratione* – whilst the poison is a substance and by implication, is evidently not a 'secret'. We might have assumed the contrary, but perhaps making a poison, although forbidden knowledge, was merely a matter of technical know-how; what was really occult was the 'secret' (or secret words) which accompanied it. However, Maddalena also used the word 'secret' to refer to a substance – in this case, a lethal potion. As she explained to the spy, 'If you dip a pen in one of my secrets … and touch a dog's or cat's nose with it you will see it die'.[70] Subsequently she explained that in order to kill a man using a letter:

> You take the written letter and place it on a table … and then with the aforementioned pen you *strego* (perform a spell) over the letter two or three times. This secret is a white thing like water which no one knows.[71]

In this case, the secret is a substance which requires the recitation of magic words to make it efficacious.

Finally, written and memorized magical formulas are referred to as secrets by the prostitute Caterina Spirilla when she describes how GioBattista had many of his illegal 'books and writings' burned by his confessor, but that 'knowing many by heart' he had written them out and kept them hidden in a bag, and 'these writings were all secrets which he had'.[72] Taken together, all the uses of the term *secreto* attributed to or used in connection with Maddalena indicate an association

---

[69] 'Mi basta l'animo di far lo morire con secreto et veleno che so far io'. Spy's testimony, ASR, TCG, Processi, vol. 114, sec. XVII, 11 July 1613, fol. c155v.

[70] 'Che bagnando una pena in un mio secreto per far sbasire una persona e tocando sotto il naso di un cane o di un gatto'. Ibid., fol. c156r.

[71] 'Ma per far morire poi un'huomo in una lettera si piglia la letera scritta et si mette stesa sopra a una tavola o a qualche cosa che sia piana et poi con detta pena si strego la sopra a detta letera doi o tre volte il qual'secreto e una cosa bianca come aqua che non si conosce'. Ibid.

[72] 'Sapindone molte alla mente… e queste scritture erano tutti secreti chi lui haveva'. Caterina's testimony, ASR, TCG, Processi, vol. 114, sec. XVII, 29 July 1613, fol. c229r-v.

between secrets and things which are hidden and magical. 'Secret' is used to denote written instructions for magical formulas; spoken magical words, an occult compound and a compound used with magical words.[73]

Conclusion

I have not found out what eventually happened to Maddalena and may never do so. The verdicts of cases before the *Governatore* were kept in separate volumes and many of these have been lost; hence the result of her trial is unclear. However, judging from her past experience, if found guilty she would probably have been handed a relatively short prison sentence. She told the spy that she had already been in prison seven or eight times, 'on account of malicious people' but that she had always managed to get out 'because she is a woman to free herself from whatever trouble she is in'.[74] Indeed, the authorities seem to have been relatively lenient with such cases. Thomas and Elizabeth Cohen, in their commentary on an earlier trial before the same tribunal that involved a woman similarly accused of casting spells, suggest that such people typically would only have been imprisoned. They point out that Italian tribunals seldom sent those found guilty of witchcraft 'to the gallows or the stake'.[75]

Regardless of the outcome of the trial, the archival trail which Maddalena's arrest has left offers us several points of entry into issues related to the compilation, circulation, making and meaning of secrets in early modern Italy. As regards secrets texts, there appear to have been marked differences in their compilation, authorship and intended publics as compared to other parts of Europe, particularly in terms of gender and social class. Previous research has shown that women were involved in the making and administration of remedies in the household and hence represented an audience for printed books of household recipes.[76] Maddalena's trial, in contrast, casts doubt on whether the making of medicines was a viable or common domestic practice in Italy. In

---

[73] For other examples of using *secreto* to indicate magical words and formulas see Barbierato, *Nella stanza dei circoli*, pp. 92 and 182.

[74] Spy's testimony, ASR, TCG, Processi, vol. 114, sec. XVII, 11 July 1613, fol. c160r. Presumably the 'malicious people' were neighbours or acquaintances who had reported her for her practices of witchcraft.

[75] Cohen and Cohen, *Words and Deeds*, p. 240.

[76] For example Cabré refers to 'compilations intended for women's use written at a single time and conceived by their authors as closed texts'. Cabré, 'Women or Healers?', p. 47. Eamon describes them as belonging to 'a large family of Italian all-purpose manuals for household use'. Eamon, *Science*, p. 127. Leong points out that 'contemporary advice literature … presented medical knowledge as essential to any early modern housewife … These views were not only prescribed but also followed'. Leong, 'Making Medicines', p. 147.

view of this, and given the commercialization of remedies in general, it makes sense that some women should have moved to occupy the lowly, 'female' fringes of the medical marketplace, making those products which were most in demand and selling them to other women who had not the time nor means to do it themselves. The circulation of Italian medicinal secrets in books, broadsheets and pamphlets may have served commercial purposes more often than domestic ones.

By looking at the use of the language of secrets in the trial, furthermore, I surmise that Maddalena was immersed in a culture in which 'secrets' had multiple and confused meanings, but still referred mainly to the esoteric. Eamon has suggested that there was a shift over the early modern period in the approach to knowledge about the manipulation of nature. Initially such knowledge was shrouded in 'secrecy' and associated with 'esoteric wisdom'. But over the sixteenth and seventeenth century he identifies a tendency to make the 'secret' knowledge of nature public and concludes that books of secrets taught people 'how to' manipulate nature's forces.[77] This association between 'secrets' and 'openness' seems to pertain chiefly to a man's world, or the world of elite practitioners.

Maddalena exhibited many similarities with this world: her knowledge of how to make oils, waters, cosmetics and even poisons resembled printed 'craft secrets'; she was open about her recipes; and she even suggested setting up an 'experience' on an animal to demonstrate her skill in poison making.[78] Yet she did not refer to her techniques or recipes as 'secrets'. Instead, her 'secrets' referred to the magical words said over her potions and to her love magic. To Maddalena, the word *secreto* had acquired none of the connotations of openness and revelation so vaunted by the 'professors' and publishers of secrets. For the most part she grounded her knowledge in a world of occult forces which few could tame. In her world, a secret was still laden with its traditional meaning, as something concealed and known only to the initiated few. It was even part of her professional identity, for she was, as she assured her potential client, a 'woman to keep this business secret' because she was a 'secret woman'.[79]

[77] Eamon, *Science*, pp. 4–5.

[78] 'Questa esperienza che io o promesso la voglio fare prima sopra un animale ma la voglio fare in presenza di questo signore vostro patrone'. Spy's testimony, ASR, TCG, Processi, vol. 114, sec. XVII, 9 July 1613, fol. c153v.

[79] 'Io so dona da tener secreto … io so dona secreta'. Spy's testimony, ASR, TCG, Processi, vol. 114, sec. XVII, 14 July 1613, fol. c162v.

PART IV
Secrets and Health

Chapter 8

Keeping Beauty Secrets in Early Modern Iberia

Montserrat Cabré*

In 1563, one of the most popular books of secrets published in Western Europe was issued in Spanish in three different editions. Alessio Piemontese's (Girolamo Ruscelli's) *Secreti* was published in Barcelona, Alcalá de Henares and Zaragoza; it was this last translation, based on the original Italian version, that became the standard in Iberia, being reprinted again in 1570 and six more times during the seventeenth century.[1] Alessio's compilation was rich in medical, alchemical and metallurgical recipes as well as in instructions for cooking and making preserves. In addition, procedures to clean and perfume the body – as well as clothes

* It is a pleasure to acknowledge the generosity of Elaine Leong and Alisha Rankin, who have been very helpful first as conference organizers and later as unflagging and careful editors. I am also indebted to various colleagues who have patiently answered my questions and enriched this paper with their suggestions: Jon Arrizabalaga, Lluís Cifuentes, José Pardo Tomás, Fernando Serrano, Caroline Wilson and finally Fernando Salmón, who witnessed its growth on a day-to-day basis. I am most grateful to Marga Vicedo and Mark Solovey, whose friendship was crucial while doing research for this paper at the Institute for the History and Philosophy of Science, University of Toronto, in May–June, 2008. Oral versions were presented at the Cambridge conference *Secrets and Knowledge*, February 2008, and at the panel *Health and the Home* organized by Sandra Cavallo at The Sixteenth Century Conference, Geneva, May 2009; I am indebted to the session commentators, Sara Pennell and David Gentilcore, as well as to their audiences for their suggestions. Material from this article was published in Montserrat Cabré, 'Los consejos para *hermosear* en el *Regalo de la vida humana* de Juan Vallés', in Juan Vallés, *Regalo de la vida humana. Estudios y transcripción*, trans. F. Serrano Larráyoz (Pamplona: Gobierno de Navarra, 2008), pp. 171–202. This research has been funded by the Spanish Ministry of Innovation and Science, project HAR2008-02867/HIST.

1 Mar Rey Bueno, 'Primeras ediciones en castellano de los libros de secretos de Alejo Piamontés', *Pecia Complutense*, 2/2 (2003): pp. 26–34. Online journal: http://www.ucm.es/BUCM/pecia/ [accessed: 30 May, 2010]. We are in need of a thorough study of the people and motives behind these three editorial processes, as well as close comparison of the printed texts. The examination of an extant copy of the Barcelona edition held at the Biblioteca de la Universitat de Barcelona (Res. 07 XVI-476) adds precision to the information given by Rey regarding the differences between the Alcalá and the Barcelona editions.

– constituted a significant presence in the text, together with an entire section devoted to cosmetics.[2]

Not long before his death in 1563 – the same year as Piemontese's debut in the Spanish press – Juan Vallés, a learned notary and royal officer of the kingdoms of Aragon, Castile and Navarre, finished an extensive compilation of household knowledge, entitled *Regalo y policía de la vida humana* (literally *Royal treatment and policy of human life*). The collection, today extant in one single manuscript copy, consisted of seven books of recipes dealing with the preparation of cosmetics, waters, oils, medicines, wines, foodstuffs and confectionary. The first three books were concerned with the external care of the body, providing instructions to improve its appearance and smell and to keep the skin clean and soft. While the second and third books were devoted to waters and oils, they also contained many recipes intended to cure internal illnesses.

This chapter is concerned with showing the emergence of cosmetics in sixteenth and early-seventeenth century Iberian practical domestic handbooks. I use the category of 'beauty secrets' to convey practical knowledge in the form of directives to prepare skin and hair treatments, dyes, depilatories, perfumes and makeup, as well as advice on how to apply them to the surfaces of the body and methods to store them for further use. Beautification recipes were copied and classified within the early modern literature of secrets and cosmetics became regularly associated with how-to texts in household settings. As in the texts by Alessio and Vallés, the word 'secret' permeated domestic guides in early modern Iberia. The term was often imbued with rich and scarcely fixed meanings that we can only infer through its contextual use. However, it never lost the basic implication that contemporary dictionaries stressed: 'Secret is everything that is under cover and kept quiet. A secret place is one unfrequented by people. A secret thing is what one entrusts to somebody'.[3]

In spite of its ambivalent – even paradoxical – uses in early modern sources, I believe that 'secret', for a variety of reasons, is an accurate concept with which to address knowledge on beautifying that appears in domestic genres. First of all, it highlights the link between cosmetics and the well-established category 'books of secrets'. Secondly, it reveals the technical character of the recipes to make beautifying products, a character that was an important feature of the epistemological status of the secrets literature.[4] Finally, it evokes a particular

[2] William Eamon, *Science and the Secrets of Nature: Books of Secrets in Medieval and Early Modern Culture* (Princeton, NJ: Princeton University Press, 1994), pp. 144–6.

[3] 'Secreto, todo lo que está encubierto y callado. Lugar secreto donde no concorre gente. Cosa secreta lo que encomienda uno a otro'. Sebastián de Covarrubias, *Tesoro de la lengua castellana o española* (Madrid: Melchor Sánchez, 1673–74), part 2, fol. 172r. Online: http://www.cervantesvirtual.com/FichaObra.html?Ref=18011 [accessed: 30 May, 2010]. The editio princeps is from 1611.

[4] For the uses of *secret* as a category of medieval and early modern knowledge, see Eamon, *Science*, pp. 15–120; Pamela O. Long, *Openess, Secrecy, Authorship. Technical*

type of knowledge of the body that, obtained through empirical methods and unexplained by causal analysis, was used, transmitted and shared by women.[5]

This chapter considers a variety of hybrid texts in which beauty secrets appear and demonstrate their importance in forging new genres of domestic handbooks. My aim here is threefold. Firstly, I will show people's active interest in keeping beauty secrets to hand. Secondly, I will illustrate the rich links connecting script and print – channels that flow in both directions. Finally, through a discussion of one manuscript – Vallés' *Regalo* – I will reveal some hidden aspects that lay behind the transformation of texts from script to print. The fate of Vallés' work has been far more modest and less visible than Alessio's successful boom in the early modern press. The text has been only recently published for the very first time, by Fernando Serrano Larráyoz.[6] And yet, it would be unfair to judge the history of the *Regalo de la vida humana* as one of non-appreciation and oblivion. The manuscript copy we know of today – probably an autograph – was highly valued by family members of later generations and by male aristocrats closely related to the Spanish court.[7] I believe that addressing the significance of 'invisible' texts such as Vallés' manuscript will extend our knowledge of books of secrets, texts which were so important in early modern culture, and explain and circumscribe their success in print.

Texts for the Accomplished Household

During the 'long sixteenth century', a wide variety of printed and manuscript handbooks emerged in Iberia for use in the 'accomplished household' (*casa cumplida*).[8] Whether in print or in manuscript, these compilations consisted of a varying group of procedures presented in the form of recipes for the elaboration and administration of cosmetics, foods and medicines; the care of clothes and leather goods; the making of preserves; the production of soaps, metals, inks and varnishes; methods to construct and reconstruct household technologies of small scale; advice for hunting and for practising the art of falconry; and for gardening

Arts and the Culture of Knowledge from Antiquity to the Renaissance (Baltimore, MD: Johns Hopkins University Press, 2001), p. 7.

5 Katharine Park, *Secrets of Women. Gender, Generation and the Origins of Human Dissection* (New York,NY: Zone Books, 2006), pp. 77–102.

6 Vallés, *Regalo*, pp. 261–756. See note 1. Serrano edited the manuscript and coordinated a team of scholars who studied different aspects of the text; a companion volume to the studies and edition contains a facsimile of the original.

7 María Itziar Zabalza Aldave, 'Estudio codicológico del Regalo de la vida humana de Juan Vallés', [MS 11160, Österreichische Nationalbibliothek, Codex Vindobonensis Palatinus], in Vallés, *Regalo*, pp. 77–122.

8 As expressed in Pedro de Sada's introduction to Vallés, *Regalo*, p. 263.

as well as farming.[9] The range of topics these manuals dealt with was extensive, as were the significant differences that the books presented in organization and lay out. As Lluís Cifuentes has recently shown, a new type of book came to be favoured by the learned elites: the miscellany encompassing a variety of texts that made apparently useful knowledge on how to care for oneself at home available to the nobility and the bourgeoisie.[10]

The increased interest in practical household books was certainly not uniquely an Iberian feature. Vernacular genres compiling practical knowledge flourished in Western Europe in the sixteenth and seventeenth centuries – from the so-called books of secrets to commonplace books, conduct books, manuals of healthcare and, especially, collections of medicinal receipts.[11] If we consider their contents and formal structure, these texts constitute a heterogeneous group that resists precise delimiting and narrow genre definitions. Nevertheless, the texts share substantial features that make it appropriate to study them as a coherent body of evidence that is historically significant. They contain concrete instructions on an extensive array of procedures to be used in the domestic setting in order to improve everyday life as well as the experience of health and well-being.

Although we lack broad systematic overviews of this literary phenomenon, these texts have been the subject of studies that evaluate their significant place in early modern culture. The Italian and English traditions have received particular attention and in-depth studies of these books are starting to show that behind the broad label of household books stands a great variety of different texts, as well as authors' and readers' aims. The printed handbooks, designed to appeal to a broad range of people as aides to ordinary life, coexisted with manuscript domestic guides in the form of recipe collections and compilations, notebooks, and so on.[12] As these notebooks demonstrate, there was a primary interest in practical

[9] Although I will refer to them incidentally, I leave aside popular texts providing clues to understanding the natural world without offering practical instructions on how to intervene in it. On this important distinction see Alison Kavey, *Books of Secrets. Natural Philosophy in England, 1550–1600* (Urbana, IL: University of Illinois Press, 2007), pp. 156–60.

[10] Lluís Cifuentes Comamala, 'La ciencia en vulgar y las élites laicas, de la Edad Media al Renacimiento', in Vallés, *Regalo*, pp. 123–48.

[11] I use here the vernacular term not only to highlight the use of non-classical languages but to mark the popular character of these texts. Mary E. Fissell, *Vernacular Bodies. The Politics of Reproduction in Early Modern England* (Oxford: Oxford University Press, 2004), p. 7.

[12] Pioneering works include Paul Slack, 'Mirrors of Health and Treasures of Poor Men: The Uses of the Vernacular Medical Literature of Tudor England', in Charles Webster (ed.), *Health, Medicine and Mortality in the Sixteenth Century* (Cambridge: Cambridge University Press, 1979), pp. 237–74; Eamon, *Science*; Ann Moss, *Printed Commonplace-Books and the Structuring of Renaissance Thought* (Oxford: Clarendon Press, 1996); Jennifer Stine, *Opening Closets: The Discovery of Household Medicine in Early Modern England* (PhD diss., Stanford University, 1996), pp. 176–215; Lynette Hunter, 'Women and

knowledge among both aristocratic and middling sort men and women. Studies of manuscript traditions are particularly revealing of the extent to which audiences were at the heart of the creation of domestic genres.[13] Authors, editors and printers worked in response to a general public demand, which prior and simultaneously to the engagement of the press, certain individuals met by writing and compiling texts for their own use and by sharing them with the textual communities to which they belonged.

The varied nature of the practical knowledge included in individual handbooks makes the frequent presence of cosmetics in these compilations, in both printed and manuscript forms, particularly significant. Procedures to beautify and care for the surfaces of the body were a regular part of women's healthcare, as attested by the flow of cosmetic texts written for women by sixteenth-century physicians.[14] However, the distinctive presence of cosmetics in domestic handbooks has not been addressed as an emergent historical issue. Jean-Lois Flandrin has written on the relationship between early modern books of secrets and beauty care, pointing

---

Domestic Medicine: Lady Experimenters, 1570–1620' in Lynette Hunter and Sarah Hutton (eds), *Women, Science and Medicine, 1500–1700* (Thrupp: Sutton Publishing, 1997), pp. 89–107; Lynette Hunter, 'Books for Daily Life: Household, Husbandry, Behavior', in John Barnard and D. F. McKenzie (eds), *The Cambridge History of the Book in Britain. Volume IV, 1557–1695* (Cambridge: Cambridge University Press, 2002), pp. 514–32; Laura L. Knoppers, 'Opening the Queen's Closet: Henrietta Maria, Elizabeth Cromwell, and the Politics of Cookery', *Renaissance Quarterly*, 60/2 (2007): pp. 464–99; Kavey, *Secrets*; and the chapters by Sandra Cavallo and Tessa Storey in this volume.

[13] Stine, *Opening Closets*, pp. 17–61; Moss, *Printed Commonplace-Books*, pp. 100–1, 106; Richard Aspin, 'Who Was Elizabeth Okeover?', *Medical History*, 44/4 (2000): pp. 531–40; Sara Pennell, 'Perfecting Practice? Women, Manuscript Recipes and Knowledge in Early Modern England', in Victoria Burke and Jonathan Gibson (eds), *Early Modern Women's Manuscript Writing: Selected Papers from the Trinity/Trent Colloquium* (Aldershot: Ashgate, 2004), pp. 237–358; Margaret Ezell, 'Domestic Papers: Manuscript Culture and Early Modern Women's Life Writing', in Michelle M. Dowd and Julie A. Eckerle (eds), *Genre and Women's Life Writing in Early Modern England* (Aldershot: Ashgate, 2007), pp. 33–48; Catherine Field, '"Many hands hands": Writing the Self in Early Modern Women's Recipe Books', in ibid., pp. 49–63; Elaine Leong and Sara Pennell, 'Recipe Collections and the Currency of Medical Knowledge in the Early Modern 'Medical Marketplace'" in Mark S. Jenner and Patrick Wallis (eds), *Medicine and the Market in England and its Colonies c. 1450–c. 1850* (Basingstoke: Palgrave Macmillan, 2007), pp. 133–52; Elaine Leong 'Making Medicines in the Early Modern Household', *Bulletin of the History of Medicine*, 82/1 (2008): pp. 145–68; Alisha Rankin, 'Duchess, Heal Thyself: Elisabeth of Rochlitz and the Patient's Perspective in Early Modern Germany', *Bulletin of the History of Medicine*, 82/1 (2008): pp. 109–44.

[14] See Alison Klairmont-Lingo, 'Santé et beauté féminines dans la France de la renaissance', in *110è Congrès national des sociétés savantes, Montpellier 1985, Hist. mod.* (Paris: Comité des travaux historiques et scientifiques, 1985), pp. 191–99; Mariacarla Gadebusch-Bondio, 'Piacevoli ragionamenti e medicina per le donne. Giovanni Marinello, medico galante del tardo Cinquecento', *Medicina nei Secoli*, 11/1 (1999): pp. 55–84.

out the importance of beauty treatments in printed compilations of secrets. Acknowledging that women were the interested party in beauty recipes, he argued that prior to industrialization, women made their own cosmetics or entrusted their production to apothecaries.[15] Edith Snook has explored seventeenth-century English women's manuscripts, showing not only their active involvement in cosmetic knowledge but also how this sphere of knowledge was part of domestic healthcare in its own right.[16] Beyond the concern of laywomen, scholars have investigated the creation of a broader appeal for beauty care services in early modern Europe, as well as male engagement with domestic medical activities.[17] Catherine Lanöe has studied the growing involvement of various male medical practitioners in the development of a cosmetic industry in Paris;[18] Sandra Cavallo has demonstrated the participation of barber-surgeons in response to the increasing demand for beauty care within the context of redefining masculinity in northern Italy.[19]

Like other European regions, Iberian texts for domestic use have started to be analyzed over the last 15 years. A significant scholarly investment has come from the growing historiographic interest in women's history and the history of everyday life.[20] However, historians of medicine sharing those concerns have also significantly contributed to defining the subject, particularly by analyzing how medical practitioners concerned themselves with providing advice to the lay for healthy living, bringing about a rich variety of genres that are currently being

---

15 Jean Louis Flandrin, 'Soins de beauté et recueils de secrets', in *Les soins de beauté. Moyen Âge – début des temps modernes. Actes du 3e colloque international de Grasse (26–28 avril 1985)* (Nice: Faculté des Lettres et Sciences Humaines, 1987), pp. 13–29.

16 Edith Snook, 'The Beautifying Part of Physic. Women's Cosmetic Practices in Early Modern England', *Journal of Women's History*, 20/3 (2008): pp. 10–33.

17 For men's involvement in household healthcare in the later seventeenth and eighteenth centuries, see Lisa Smith, 'The Relative Duties of a Man: Domestic Medicine in England and France, ca. 1670–1740', *Journal of Family History*, 31/3 (2006): pp. 237–56.

18 Catherine Lanoë, *Le poudre et le fard. Une histoire des cosmétiques de la Renaissance aux Lumières* (Seyssel sur le Rhône: Champ Vallon, 2008).

19 Sandra Cavallo, *Artisans of the Body in Early Modern Italy. Identities, Families and Masculinities* (Manchester: Manchester University Press, 2007), pp. 38–63.

20 María de los Ángeles Pérez Samper, 'Los recetarios de mujeres y para mujeres. Sobre la conservación y transmisión de los saberes domésticos en la época moderna', *Cuadernos de Historia Moderna,* 19 (1997): pp. 121–54; eadem, 'Las mujeres y la organización de la vida doméstica: De cocineras a escritoras y de lectoras a cocineras', in Tomás A. Mantecón (ed.), *Bajtin y la historia de la cultura popular* (Santander: Publican, 2008), pp. 33–69; Monica Bolufer, 'Medicine and the *Querelle des Femmes* in Early Modern Spain', *Medical History. Supplements*, 29 (2009): pp. 86–106, especially pp. 95–102; María Ángeles Ortego Agustín, 'Discursos y prácticas sobre el cuerpo y la higiene en la edad moderna', *Cuadernos de Historia Moderna. Anejos*, 8 (2009): pp. 67–92.

singled out.[21] My aim in this chapter is to contribute to these trends by addressing two aspects that, as I will argue, are intertwined in early modern Iberia: on the one hand, the process of formation of domestic genres of self-care; on the other, the importance of cosmetics within this literature for household use that came to be favoured by a wide spectrum of the social scale.

Appropriating Cosmetic Recipes

It is not uncommon to find cosmetics treated as a given fact of early modern culture, requiring light to be thrown on either the heated discussions on its legitimate moral practice or on its particular development within the history of a growing healthcare industry. Debates on the proper and improper use of cosmetics have attracted the most attention, particularly from art historians and literary scholars. Studies on the moral, religious, aesthetic and political judgments in favour of or against cosmetics have placed the issue in the open arena of early modern cultural studies.[22] Nevertheless, the practice of cosmetics itself is often portrayed as an ahistorical sphere of body care. Accounts of its *emergence* – rather than its presence – in early modern texts do not abound, and beauty recipes – and the earlier traditions they belong to – are less frequently the object of historical study than other types or recipes they often accompany.

Cosmetics – and perfumes – had a visible place within medicine in medieval Latin Europe. The sphere of practices understood as *decoratio* and *ornatus* offered techniques for caring and modifying the surfaces of the body: the cleaning and softening of the skin, the hygiene of the mouth, the care and colouring of the hair of the head and beard, the elimination of lice, the depilation of unwanted body hair and the treatment of all kinds of skin imperfections. The technical literature of *Antidotaria* noted to a significant extent the cosmetic properties of the

21 Studies include José María López Piñero, *El vanquete de nobles cavalleros (1530), de Luis Lobera de Ávila y la higiene individual del siglo XVI* (Madrid: Ministerio de Sanidad y Consumo, 1991); María José Ruiz Somavilla *El cuerpo limpio. Análisis de las prácticas higiénicas en la España del mundo moderno* (Málaga: Universidad de Málaga, 1993) and eadem, 'Las normas de higiene y los consejos de carácter moral en la práctica médica de los siglos XVI y XVII' *Dynamis*, 22 (2002): pp. 235–50; Montserrat Cabré, 'From a Master to a Laywoman: A Feminine Manual of Self-Help' *Dynamis*, 20 (2000): pp. 371–93; Luis García Ballester, *La búsqueda de la salud. Sanadores y enfermos en la España medieval* (Barcelona: Península, 2001), pp. 293–410; Lluís Cifuentes, *La ciència en català a l'edat mitjana i el renaixement* (Barcelona: Universitat de Barcelona, 2002), pp. 96–112; Lluís Alcanyís, *Regiment preservatiu e curatiu de la pestilència*, edited by Jon Arrizabalaga (Barcelona: Barcino, 2008), as well as the essays in Vallés, *Regalo.*

22 For an original study on the debates on women's painting (encompassing pictures and faces) in regards to England, Italy and France, see Patricia Phillippy, *Painting Women. Cosmetics, Canvases and Early Modern Culture* (Baltimore, MD: Johns Hopkins University Press, 2006).

samples included; this amounted to 20 per cent in the case of the popular *Circa instans* attributed to Matthaeus Platearius. Most of the practical medical treatises *de capite ad calcem* incorporated sections devoted to these topics and surgical literature included sections on cosmetic procedures.[23] Many of these treatments were ungendered, but some were specifically intended for men, particularly those regarding the care of the beard. However, most cosmetic proceedings were associated with women, either specified by the recipes or linked by the actual practices.[24] In fact, Salernitan medical culture was the origin of a genre of texts devoted to women's cosmetics, usually associated with other texts on female healthcare concerns.[25] By the end of the thirteenth century, texts addressed to laywomen with a significant cosmetic content had started to flourish in different vernacular languages.[26] Although medical handbooks included cosmetic treatments for men, a male counterpart of this feminine genre does not seem to have arisen in Western Europe. While still keeping their place in learned Latin medicine and surgery, during the late Middle Ages beautifying recipes started to become part of the lay written culture, particularly those addressed to women.

As evidence of early modern women's growing interest to in possessing beautifying recipes, texts both compiled for and compiled by women steadily increased in fifteenth- and sixteenth-century Iberia.[27] In this period, cosmetic recipes are found in a variety of manuscript contexts: from extensive compilations containing hundreds of items to small collections of only a handful, or even exchanges of single individual pieces which eventually may have been compiled in a collection. Compilations differ from each other not only in size but also in the way they privilege certain topics over others. For instance, in an account book that the lawyer Bernat Sala ended up sharing with his second wife Agnès Safont – but that had been started by his first wife – Isabel de Seix, one out of the nine of recipes deals with beautifying: a water to clarify the skin of the face. The rest are devoted

23 Walton O. Schalick, 'The Face Behind the Mask: 13th- and 14th- Century European Medical Cosmetology and Physiognomy', in Yasuo Otsuka, Shizu Sakai and Shigehisa Kuriyama (eds), *Medicine and the History of the Body* (Tokyo: Ishiyaku EuroAmerica, 1999), pp. 295–311; Laurence Moulinier-Brogi, 'Esthétique et soins du corps dans les traités médicaux latins à la fin du Moyen Âge', *Médiévales*, 46 (2004): pp. 55–72; Michael McVaugh, *The Rational Surgery of the Middle Ages* (Firenze: Sismel- Edizioni del Galluzzo, 2006), pp. 215–29.

24 Montserrat Cabré, 'Beautiful Bodies', in Linda Kalof (ed.), *A Cultural History of the Body in the Middle Ages* (Oxford: Berg, 2010), pp. 127–48.

25 Monica H. Green, *The Trotula: A Medieval Compendium of Women's Medicine* (Philadelphia, PA: University of Pennsylvania Press, 2001).

26 Monica H. Green, 'The Possibilities of Literacy and the Limits of Reading: Women and the Gendering of Medical Literacy', in eadem, *Women's Healthcare in the Medieval West* (Aldershot: Ashgate, 2000), Essay VII, pp. 32–7 and 49–76; Cabré, 'From a Master'.

27 Montserrat Cabré, 'Women or Healers? Household Practices and the Categories of Health Care in Late Medieval Iberia', *Bulletin of the History of Medicine*, 82/1 (2008): pp. 18–51 (pp. 36–50).

to fruit preserves, and besides a systematic register of the loans of money lent and borrowed, the notebook records certain purchases as well as poems. Interestingly, as is the case with other similar texts, this small group of recipes is written both in Catalan and Spanish. It is unclear which individuals actually wrote this handful of recipes but, together with other stylistic differences, the different languages may indicate a different provenance of the recipes recorded in the manuscript.[28]

Texts that record miscellaneous personal annotations, such as accounts, purchases and certain family affairs, often contain recipes; their significance is currently being re-evaluated by scholars working on the so-called 'domestic writings' or 'ego documents'.[29] In this context, when found as part of small groups of recipes, cosmetics are generally associated with cooking recipes and/or with instructions on how to make food preserves. However, if they are present in more extensive texts, the scope of the topics addressed seems to widen. The collection compiled by Juan Vázquez de Mármol, secretary to the Consejo de Castilla under Kings Philip II and Philip III, is a good instance of this trend. A royal priest and a humanist who worked as a general book reviser for Philip II, Vázquez de Mármol wrote a long recipe book containing precise instructions to produce inks and dyes, to remove difficult stains, to make preparations to improve memory, to construct watches, to make glues, to kill fleas and other types of insects, to prepare antidotes for poisons as well as to make medicines for a great variety of common ailments – such as ear and stomach pains, headaches, and so on – among other recipes on smaller-scale how-to household technologies. Only a few recipes in this manuscript are concerned with food, but over 50 deal with the care of the hair and body surfaces as well as the improvement of the smell of the body and breath.[30] As its title indicates, the *Receptario sacado de Don Alexio Piemontés y de*

---

28 *Llibre de memòries de Bernat Sala, Isabel de Seix i Agnès Safont, ca.* 1605–1628. Barcelona, Arxiu de la Corona d'Aragó, Monacals d'Hisenda, vol. 3401. A general description in *Repertori de manuscrits catalans (1474–1620)* (Barcelona: Institut d'Estudis Catalans, 2003), vol. 3, pp. 337–8. The recipe collection is in fols 2r-6v and 42v.

29 Sylvie Mouysset, 'Maux dits, maux écrits: Du soin de soi à l›attention aux autres dans les écrits du for privé français, XVème–XVIIIème siècle', in Antonio Castillo and Verónica Sierra (eds), *El legado de Mnemosyne. Las escrituras del yo a través del tiempo* (Gijón: Trea, 2007), pp. 17–37; eadem, *Papiers de famille. Introduction a l'étude des livres de raison (France, Xve–XIXe Siècle)* (Rennes: Presses Universitaires de Rennes, 2007), p. 303; Daniel Piñol, 'Salud, dinero y amor sobre el papel: Usos domésticos de la escritura en Reus (Siglos XVIII–XIX)', in Castillo and Sierra (eds), *El legado de Mnemosyne*, pp. 39–54; Carmen Rubalcaba, *Entre las calles vivas de las palabras. Prácticas de cultura escrita en el siglo XIX* (Gijón: Trea, 2006); Rosa María Blasco and Carmen Rubalcaba, 'Las escrituras del yo en los libros de cuentas de Pedro Jado', in Castillo and Sierra (eds), *El legado de Mnemosyne*, pp. 55–74.

30 Madrid, Biblioteca Nacional de España (thereafter, BNE) MS 9226, fols 145r–225v. The recipe book starts with an index of recipes according to a different, individual numbering of its pages, from 1 to 148, identifying it as a separate text. It was bound by its compiler together with other personal texts containing reading notes, registers

otros autores y de otras muchas receptas que me han dado (*Book of recipes culled from Alessio Piemontese as well as from many other authors and recipes that have been given to me*) contains recipes copied from the printed *Secreti*, which the compiler identifies individually, recipe by recipe, with his source. However, Vázquez de Mármol may have used an Italian edition instead of (or in addition to) a Spanish translation since it contains recipes written in Italian.[31] Although the compiler acknowledges other printed books and medical men as his sources, most of the beauty recipes he writes seem to derive from Alessio's *Secreti*. And incidentally, he adds commentary on his source: in a recipe to help hair grow, for instance, he notes that Alessio does not make it clear which of the ingredients he has listed for the recipe has to be fried – showing his commitment to understand and judge what he writes.[32]

Another potent example of the extent to which beauty recipes came to be seen as an inextricable part of domestic knowledge is a copy of the 1516 edition of the Salernitan *Regimen sanitatis* with commentary, which calls into question textual genres and categories of knowledge as well as notions of authorship and readership.[33] Published in Lyon and today held at the Biblioteca General de Navarra in Pamplona, this Latin regimen of health, like many printed and manuscript books, contains a few marginal anonymous notes. Most interestingly, following the end of the printed text is a collection of over 40 household recipes written by two anonymous sixteenth-century hands in Latin, Spanish and Catalan. About 30 of the entries are recipes to care for and modify body surfaces – the cleaning and care of the skin, including ulcers on the penis – and to look after and dye the hair, to care for teeth and gums and to eliminate smelly feet and bad body odours. The rest are to improve memory, to have a good sleep, to make ink, to

of lent and borrowed books, historiographical and hagiographical accounts, a Spanish-Flemish dictionary as well as texts regarding the art of printing. One precise date appears in the compilation, when Vázquez de Mármol writes that on January 20, 1596, he copied a recipe for a purge by master Guillem de Mallorca that he had found at the margin of a compilation of Spanish law that he borrowed from a friend, fol. 199r. On aspects of this manuscript other than the recipe collection, see Fernando Bouza, *Corre manuscrito: Una historia cultural del Siglo de Oro* (Madrid: Marcial Pons, 2001), pp. 50–1. For an analysis on the production of miscellaneous manuscripts and annotating practices in early modern Iberia, see Fernando Bouza, *Communication, Knowledge and Memory in Early Modern Spain*, translated by Sonia López and Michael Agnew (Philadelphia. PA: University of Pennsylvania Press, 2004), pp. 39–55.

[31] BNE, MS 9226, fol. 169r.

[32] Ibid., fol. 186r.

[33] *Regimen sanitatis salerni: Accurate castigatum adiecta tabula in calce libri hactenus non impressa* (Lyon: Jacobum Myt, 1516). Pamplona, Biblioteca General de Navarra, Fondo Antiguo, NA-BGN, 12-2/39. The collection of recipes proper starts at the end of the alphabetical table and extends to seven unnumbered folia (14 pages). I am indebted to Fernando Serrano Larráyoz and Margarita Velasco Garro for generously bringing this source to my attention.

paint glass, one for ear pain, two for hunting, for treating stones in the urine and one to help expel the placenta after childbirth. This collection again demonstrates a widespread interest for beauty recipes – which were by far the largest of the additions – and also the commitment of readers of printed books 'to complete' the texts they owned according to their wishes, literally producing new books for their own use.

Regularly culled from people and books – both printed as well as handwritten, beauty recipes became part of the textual culture of the household in sixteenth-century Iberia. Two significant features emerge from the documented compilation practices: 1) the extent to which people with no direct connection with the healthcare occupations were actively involved in the process of written appropriation; 2) beauty recipes were associated with other types of practical knowledge useful to have to hand in the domestic setting. The diversity of genres where single recipes and extensive collections surface attest to the appropriation of beauty secrets by women and men who associated cosmetics with the everyday, domestic sphere of care.

Between Script and Print: The *Regalo de la vida humana*

The *Regalo y policía de la vida humana*, a sizable collection of medical and culinary recipes as well as precise instructions on how to make wines, vinegars, oils, waters and food preserves, included a significant amount of beauty procedures. Besides the embellishing properties of many of the oils and waters contained in Books Two and Three, the text started with a section specifically devoted to beautifying. In contrast with the handbooks containing practical domestic knowledge discussed above, Juan Vallés' work not only functioned to serve his own endeavours as a learned gentleman but also aimed to help his readers accomplish a wide range of household activities.

This social or public character of a text that hoped to reach an interested audience was enhanced after Vallés' death in 1563. His grandson, Pedro de Sada y Vallés (+1615), edited the manuscript with the explicit intention of publishing it under his grandfather's name. In the form of various explicit layers, the manuscript of the *Regalo* bears witness to a long process of preparation of a compilation for the press. According to the codicological study by María Itziar Zabalza, Vallés wrote a first version which was later cleaned up by himself, perhaps with the help of a scribe. The extant copy is a duplicate of this polished version, containing numerous crossing outs and additions in the margins by Vallés' hand. Around 50 years later Pedro de Sada – who may have inherited his grandfather's library – revised the text, adding some recipes and comments, numbering the pages again and writing an index.[34]

34 Zabalza, 'Estudio codicológico'.

Following his grandfather's footsteps, Pedro de Sada was a lawyer and a royal officer of the Kingdom of Navarre. In the opening epistle, addressed 'to the reader', written circa 1610, he introduced himself as an official receiver (*síndico*) of the Kingdom of Navarre and as a consultant of the Holy Office of the Inquisition.[35] While justifying his own editorial decision to bring about the printing of the text, he stated regarding the work of Vallés:

> And thus, as someone [Juan Vallés] who had always professed much curiosity, he took that to the good point [of writing this book]. Until the present day, people who have seen it have regarded it highly, as one of the better and more important items that may be hold in a principal house. It seemed to me, then, that I could not but communicate it to everyone; otherwise I would offend very much the good will of the writer as well as causing universal harm to the good of all. Therefore, I determined to try to print it so that after death his good wish to be profitable to all will not be lost by my negligence – or the overall gratitude that he deserves for such good concern.[36]

For reasons we cannot specify, Pedro de Sada died without success in this printing endeavour. However, he was familiar with the printing business since together with a fellow colleague he published a compilation of the laws of the Navarrese kingdom in 1614. Nonetheless, his editorial work gave renewed value to the *Regalo de la vida humana* and eventually facilitated its long-term transmission. The only extant copy of the text was part of the personal library of Pedro de Navarra, Marquis of Cábrega, who might have obtained it directly from Pedro de Sada – either as a gift or by acquisition. He was a significant figure at the Spanish court, having served as a steward (*mayordomo*) to Queen Mariana of Austria and as officer (*gentilhombre de boca*) to King Philip IV, specializing in the supply, conservation and preparation of the food of his house. In 1674, the Austrian ambassador in Madrid bought all items of his library for the collection of the Imperial Court Library of the Habsburg dynasty, including the manuscript of the *Regalo*. It was then rated as one of the most expensive of the cluster – a clear sign of the extent of the appreciation of its

---

35 Fernando Serrano Larráyoz, 'Juan Vallés (c. 1496–1563): Vida y obra (Regalo de la vida humana) de un humanista navarro de la primera mitad del siglo XVI', in Vallés, *Regalo*, pp. 18–75 (pp. 61–4).

36 'Y assí, como quien professó siempre mucha curiosidad, traxo ésta a tan buen punto que, al parecer de quantos hasta hoy este libro han visto, ha sido estimado por una de las mejores y más importantes cossas que en una cassa principal pueden tenerse. Paresciéndome pues, que según esto yo no podía dexar de comunicarlo a todos sino haziendo mucho agravio a la buena intención del que lo trabajó y daño universal al bien común de la gente, determiné procurar se imprimiesse sólo porque el buen desseo que en vida él tubo de aprovechar a todos no perdiesse por mi descuydo en su muerte las gracias que generalmente todos le deben por tan buen cuydado …'. Vallés, *Regalo*, p. 263.

contents, particularly considering that it is not an illustrated text. Since then, the copy has been held in this collection, which today belongs to the Austrian National Library in Vienna.[37]

If his grandson was familiar with the printing circles, Juan Vallés was already a skilled author when he compiled the *Regalo de la vida humana*. His first work, today unknown, was called *Flores de cirugía y medicina* (*Flowers of surgery and medicine*), but his most acknowledged contribution was the *Libro de acetrería y montería* (*Book of falconry and hunting*), a comprehensive encyclopedia dated in 1556. The treatise was considered to be a Spanish landmark of its genre and gave Vallés a reputation as an author, but although he had obtained a printing licence for it, he did not see it released by a press.[38]

With the *Regalo de la vida humana*, while it is uncertain whether Vallés had undertaken such a concrete move towards publication, Pedro de Sada writes that his grandfather had planned his text 'to be profitable to all' and that the text was highly appreciated by those who 'had seen it', suggesting an initial intention towards printed publication.[39] Moreover, in reference to the final section of the book, de Sada explains that, although its inclusion is due to his own editorial decision Vallés 'wanted to print it separately' but had left it unfinished. De Sada justified its inclusion as it was 'a forest of diverse recipes in which a great variety of curiosities may be found'.[40]

Ultimately, we do not know why neither Vallés nor de Sada succeeded in their attempts to print the *Regalo*. Contemporary manuscript practices and the printing of translated books of recipes for lay readers suggest that the demand for recipes for domestic use was broadening.[41] However, as far as we can tell, the *Regalo* stands as the first early modern Spanish attempt to print a recipe

37 Fernando Serrano, 'Juan Vallés (c. 1496–1563): Vida y obra', pp. 64–5.

38 For an informed biobibliographical study together with the full text, Jose Manuel Fradejas, 'Juan Vallés, Libro de acetrería y montería', *Archivo Iberoamericano de Cetrería*, 3 (September 2007). Online: http://www.aic.uva.es/clasicos/valles/valles-intro.html [accessed: 30 May, 2010].

39 On the different means of 'publishing' in early modern Spain (of which printing was but one of a variety), see Fernando Bouza, *Papeles y opinión: Políticas de publicación en el siglo de oro* (Madrid: CSIC, 2008).

40 'En el octavo y último se pone una silva de varias recetas, en las quales se hallará una grande variedad de curiosidades no menos de provecho que de gusto …'. de Sada, 'Introduction', in Vallés, *Regalo*, p. 264.

41 Useful evaluations of early medical and scientific printing, Bertha Gutiérrez Rodilla, 'Los textos médicos romances en el Renacimiento castellano', in J.L. García Hourcade and J.M. Moreno Yuste (coords), *Andrés Laguna. Humanismo, ciencia y política en la Europa Renacentista* (Valladolid: Junta de Castilla y León, 2001), pp. 529–38; Jon Arrizabalaga, 'El libro científico en la primera imprenta castellana (1485–1520)', in Luis García Ballester (dir.) *Historia de la ciencia y de la técnica en la Corona de Castilla, Edad media II* (4 vols; Valladolid: Junta de Castilla y León, 2002), vol. 2, pp. 619–49, and José Pardo Tomás, 'La difusión de la información científica y técnica', in José María López

book of its kind. Consequently, at the end of the sixteenth century, it was not only in England that the medieval traditions of household knowledge were being eclipsed by new texts. Similar trends, albeit timidly, were also occurring in Iberia.[42]

Vallés' decision to title his text *Regalo de la vida humana* was original. If other popular books containing recipes or practical knowledge at that time were entitled *Treasure*, *Mirror*, *Garden, Forest* (Silva) or simply *Handbook*, he very precisely distinguished his work with a title that disclosed the nature of its aims. *Regalo* is a Spanish word that came to mean 'gift', etymologically deriving from the Latin word *rex, regis*. In the early modern period it meant '[the ability to offer] oneself a royal treat and to give to oneself the delights that kings may have at will'.[43] The metaphor behind Vallés' title strongly evokes the idea of body pleasure or 'delight'; a promise which, elsewhere in Europe, other recipe books of this kind were starting to offer to their intended, particularly female, readers. We see examples such as *Delightes for Ladies* by Hugh Platt (1600?), *A True Gentlewomen's Delight* associated with Elizabeth Grey, Countess of Kent (1653), *A Queen's Delight* (1668) and *The Accomplish'd Lady's Delight* (1675).[44] Vallés' choice of title is an early instance of a tradition that conceptualized a diversity of procedures leading to beautifying, eating, drinking and general feeling of well-being, as actions potentially pleasing the body. Or, in Vallés' words, as potentially delightful to human living.

By locating his book under the umbrella of 'royal treatment', Vallés was literally projecting a courtly culture that promoted individual bodily practices as the axis of social relations and human interaction.[45] Alison Klairmont-Lingo has pointed out that within this context beauty, alongside table manners and refraining from public spitting, was conceived as an important asset.[46] De Sada himself writes that these recipes promising beauty – the knowledge of how to embellish oneself – came to be a significant part of 'those curiosities that for the service and good government and policy of an accomplished household might

Piñero (dir.), *Historia de la ciencia y de la técnica en la Corona de Castilla, siglos XVI y XVII* (4 vols; Valladolid: Junta de Castilla y León, 2002), vol. 3, pp. 189–217.

42 Stine, *Opening Closets*, pp. 90–92; Hunter, 'Books for Daily Life'.

43 '[Capacidad de ofrecerse] un trato real y regalarse las delicias que los reyes pueden tener a rege, Covarrubias, *Tesoro*'. 2nd part, fol. 157r.

44 The publishing history of these works and their titles is complex, see Hunter, 'Books for Daily Life', pp. 521–2 and pp. 528–30. For analysis of these texts, see Snook, 'The Beautifying Part of Physik'; Knoppers, 'Opening the Queen's Closet'; Kavey, *Secrets*, pp. 95–124 and pp. 156–60. A comparative history of changing patterns of conceptualizing domestic knowledge, as well as what was considered a delight and for whom, remains to be undertaken.

45 Norbert Elias, *The Civilizing Process. The Development of Manners*, trans. Edmund Jephcott (New York, NY: Urizen Books, 1978).

46 Klairmont-Lingo, 'Santé et beauté féminines'.

be necessary'.[47] Practical knowledge on beautifying belonged to the sphere of good domestic management.

In addition to his singular conceptualization of the care of the body's surfaces as part of the highest possible indulgence, Vallés was also original in the way he proffered his material. Although he copied literally and extensively from earlier compilations without any acknowledgement, he organized his topics in new and important ways. In contrast to other household guides, Vallés began his compilation with issues dealing with beautifying (*hermosear*) and presented the external care of the body as the first theme to be addressed. Vallés explains this deliberate ordering of information in chapter 2 of Book One:

> Having to write in this treatise about the things regarding the 'royal treatment' of this life, it is fair that before talking about preserves, electuaries and other delights and treats which one may enjoy *on the inside*, we offer cosmetic and beautifying things to embellish the person's *outside*.[48]

Vallés use of *inside* and *outside* as organizing principles that give coherence to the structuring of knowledge – and the precedence of the latter – is novel in self-care literature and none of the earlier texts from which Vallés copied follow this divide nor prioritize the external appearance of the body.

Vallés adoption of the binary terms inside/outside as the organizational principal of his work is part of a larger cultural recognition that the human body possessed an interior and an exterior part. This notion seems to have become stronger in the second half of the sixteenth century, both for healthcare practitioners and for the general populus. Maria Carla Gadebusch-Bondio has pointed out an increasing medical concern with the skin, '*la carne di fuori*' in the words of Giovanni Marinello, from an intellectual and a practical point of view.[49] The visible quality of the body – of its public presentation and display – came to be intensified not only by the dissemination of more sophisticated regimens but also by new anatomical works that often deployed vividly the skin as the container and concealer of the body's interior. Earlier uses of the metaphor by learned surgeons suggest that the conceptual divide between an inside and an outside of the body may well have grown together with the intellectual and occupational separation of medicine and

---

47 '... aquellas curiosidades que para el servicio de una persona y buen gobierno y policía de una cassa cumplida, podían ser necesarias.' de Sada 'Introduction'. Vallés, *Regalo*, p. 263.

48 My emphasis. 'Haviendo de escribir en este tratado de las cosas que tocan al regalo desta vida, justo es que antes que hablemos de las conservas, electuarios y otras delicadezas y regalos con que se regala la \per/sona por la parte de dentro la regalemos con las cosas que la affeitan y hermosean por la parte de afuera ...'. Vallés, *Regalo*, Book 1, Ch 2, p. 279.

49 Maria Carla Gadebusch-Bondio, 'La carne di fuori. Discorsi medici sulla natura e l'estetica della pelle nel '500', *Micrologus*, 13 (2005): pp. 537–70.

surgery.[50] Recently, Michael McVaugh has explained how thirteenth- and early fourteenth-century learned surgeons struggled to define their identity by integrating Arabic medical knowledge and distillation and sublimation techniques into their craft. These interconnected developments broadened their abilities to modify the appearance of the body and opened their occupation to a sphere of practice with which they did not feel too comfortable: that of embellishing the body surfaces. However, they acknowledged a demand for those treatments – particularly by women – and responded to it, albeit with caution.[51] Eventually, barbers and apothecaries took on some of the inconvenient concerns, and surgeons expanded their practice towards new arenas regarding the treatment of women.[52] Running in parallel with the realignment and reorganization of medical practitioners grew an active interest among laypeople in collecting recipes to care 'for the outside'.

An Archaeology of Beauty Secrets: Gendering Texts and Audiences

The *Regalo de la vida humana* also differs from earlier texts conceived for household use in its intended audience. Late medieval vernacular texts tailor-made for their lay commissioners were often gendered by a direct address in their introductory prefaces, and book-length recipe collections that preceded the *Regalo* had marked their audiences' gender by their titles. In contrast to earlier Iberian compilations which were often clearly associated either with one or the other sex, de Sada, in an introduction addressed to an ungendered 'someone' or 'any person' highlighted the intent of the text to serve a space – the 'accomplished household' – rather than targeting any specific group of people as its readership.[53] This echoes the broad audience sought by contemporary printed texts such as the Spanish translation of Alessio's *Secreti*, which includes women explicitly.[54]

---

50 Marie-Christine Pouchelle, *The Body and Surgery in the Middle Ages* (Cambridge: Polity Press, 1990), pp. 125–60.

51 McVaugh, *Rational Surgery*, pp. 181–229.

52 For these developments in Italy and France, see Cavallo, *Artisans of the Body*; Lanöe, *Le Poudre et le Fard*. See also Monica Green, *Making Women's Medicine Masculine. The Rise of Male Authority in Pre-modern Gynaecology* (Oxford: Oxford University Press, 2008).

53 Together with 'someone' and 'any person', the opening letter mentions once 'men' as the book's addressees; however, the general tone of the address makes clear that it is intended at a broad audience that includes both sexes. De Sada, 'Introduction' in Vallés *Regalo*, p. 263.

54 'Del traductor al lector: … por ser médico me afficioné a estos secretos, que me parecieron muy buenos, sabiendo usar dellos a su tiempo y razón: lo otro porque libro tan curioso para cavalleros, damas, gentiles hombres y galanes y para otras muchas gentes …'. Alessio Piemontese, *Seis libros de secretos llenos de maravillosa diferencia de cosas, traduzidos de lengua latina en castellana* (Barcelona: Claude Bornat, 1563), fol. 4r.

A comparison of the *Regalo* and two thematically similar recipe collections from the late-fifteenth/early-sixteenth century shows that Vallés and de Sada's decisions to address his work to a broad general audience is closely linked to the compilation strategies employed in the creation of the text. The *Manual de mugeres en el qual se contienen muchas y diversas reçeutas muy buenas* (thereafter *Women's Handbook*) and *Vergel de señores en el qual se muestran a hazer con mucha excelençia todas las conservas, electuarios, confituras, turrones y otras cosas de açúcar y miel* (thereafter, *Garden of Gentlemen*) are two anonymous manuscripts dealing with a set of themes very similar to the *Regalo de la vida humana*.[55] They include recipes to make food preserves, meals, waters, perfumes and a significant number of procedures to produce beautifying products such as skin cleaners and softeners, depilatories, hair protectors and hair dyes as well as powders to whiten and colour the face. Roughly, the recipes seeking to offer embellishing instructions amount to more than a third of their contents in both cases. However, when closely considered, they turn out to be different types of texts and, I contend, they belong to two distinct traditions.

The *Women's Handbook* is the earliest Iberian example of a genre that I have called elsewhere '*open* recipe books', a type of recipe collections that originates out of women's annotating practices of their own domestic endeavours.[56] In this collection, the recipes included are fairly simple; the text is structurally disorganized and in all but one case it only offers one recipe for each of the aims considered. There is no voice of a compiler unifying the pieces that constitute the text and references to provenance of the knowledge written – either to individuals, authors or books – are totally absent. The *Women's Handbook* now exists in a single copy written in a practised uniform hand; however, the manuscript bears signs which suggest that it was the edited result of an individual or group enterprise of gradual appropriation of recipes for one's own use. This is most visible in the different organizational strategies evident in the information retrieval devices and in the main body of the text itself. While the manuscript presents a table of contents and a list of all the recipes organized in seven groups of topics, the order of the recipes in the text does not follow this or any other thematic criteria. Recipes written consecutively within the body of the text are assigned to different thematic groups in the table; they are also identified in different ways, some with a rubric describing the type of remedy (*ungüento cetrino*), others by the indication of the procedure (*para quitar manchas*) or by a combination (*unçión para los pechos de las mugeres paridas*).[57] The unsystematic way of organizing as well as the diverse phrasing of the recipes themselves suggest that the compilation is the result of a growing accretion of recipes that came from a variety of sources, finally

---

55 *Manual de mugeres en el qual se contienen muchas y diversas reçeutas muy buenas*, ed. Alicia Martínez Crespo (Salamanca: Universidad de Salamanca, 1995); *Vergel de señores*, BNE, MS 8565, VI + 239 fols.

56 See Montserrat Cabré, 'Women or Healers', pp. 43–50.

57 *Manual de mugeres*, pp. 30–35.

materializing in the text we know today. It is the written, edited result of women's knowledge – women's secrets.

Significantly, although never explicitly gendered, both the intended users and recipients of the practical knowledge contained in the *Women's Handbook* are women. It is not just a text compiled *by* them and *for* them, but, in regards to beauty, it is knowledge to be applied *on* them. This point becomes particularly apparent when we compare *Women's Handbook* with a contemporary collection that explicitly addressed men. Like the *Women's Handbook*, the anonymous *Garden of Gentlemen* was intended for the use of lay people in a domestic setting. However, it differs from the *Women's Handbook* in significant ways. Firstly, it is a lengthier collection organized thematically in five books, each divided into individually identified chapters. Secondly, while the only known copy does not include a prologue, the voice of the anonymous compiler is evident throughout the text and ascribes recipes to authors such as Galen and Avicenna.[58] Thirdly, instead of offering a single recipe for each intended procedure, as in the *Women's Handbook*, it provides a handful of options, presented consecutively, for each desired action. Finally, other than the title, the *Garden of Gentlemen* presents other internal signs indicating the gender of the intended audience. For example, four chapters are dedicated to the care of the male body surfaces and contain recipes for the hair and beard – for avoiding and curing baldness – as well as preparations to prevent the greying of hair and, if that recipe was unsuccessful, for dyeing grey hair of the scalp and face.[59] These Iberian recipes, offering precise instructions on how to care for and embellish beards are the practical counterpart to the growing concern for the male beard as a sign of masculinity that has been identified in literary and artistic sources of Renaissance Italy and England.[60]

The *Women's Handbook* and the *Garden of Gentlemen* are the earliest Iberian household compilations that bring together in a single unified text a certain set of topics: beauty, culinary recipes, recipes for food preservation, recipes for common illnesses.[61] Despite their overlapping thematic interests, these two collections

---

58 'Asimismo, Avicena alaba muy mucho, para hermosear y blanquear la cara e para quitarle qualquier quemadura del sol e aire, usar un ungüento que se haze desta manera'. 'Vergel de señores', BNE, MS 8565, fol. 141v.

59 'Cómo se conservarán los cabellos así de la cabeça como de la barba que no se hagan canos' 'Cómo se harán negros los cabellos canos y blancos así de la cabeça como de la barba'; 'Para hazer nacer y crecer los cabellos o pelos que se cayeron de la cabeça o barba'. Ibid., fols 130r–132r; 'De los xaboncillos de olor que se azen así para la barba como para las manos'. Ibid. fol. 216r.

60 Will Fisher, 'The Renaissance Beard: Masculinity in Early Modern England', *Renaissance Quarterly*, 54 (2001): pp. 155–87; Douglas Biow, 'The Beard in Sixteenth-Century Italy', in Julia L. Hairston and Walter Stephens (eds), *The Body in Early Modern Italy* (Baltimore, MD: Johns Hopkins University Press, 2010), pp. 176–94.

61 Cifuentes, 'La ciencia en vulgar y las élites laicas'.

clearly developed from different traditions: one associated with women's practices as well as their own social networks and the other connected to the medical and culinary knowledge sustained by learned men. These genealogies of knowledge production, evaluation and transmission were not opaque to each other; in fact, the male medieval tradition readily acknowledges women's practices as the origin of certain recipes.[62] It is this second tradition, exemplified by the *Garden of Gentlemen*, that encompasses the needs of the male body as well as those of the female. It also includes advice on the art of falconry, a male activity at the time.

The *Regalo de la vida humana* clearly belongs to the male genealogy of knowledge. In fact, in compiling the text, Juan Vallés copied word-for-word long sections from the *Garden of Gentlemen* without acknowledgement.[63] However, unlike the *Women's Handbook* or the *Garden of Gentlemen*, the *Regalo* is the result of the textual effort of a self-identified author who addressed his work to a broad audience. Pedro de Sada described his grandfather's desire to 'bring together and put in order' all the things that would be 'profitable to everyone and for common service', useful because 'neither in health may one do without them nor they can be excused in case of illness'.[64] Vallés, and later de Sada, both extended the *Garden of Gentlemen* with additional recipes. For instance, of the 17 recipes in the section on how to dye hair blond only four recipes were from the *Garden of Gentlemen*. De Sada, in turn, also added a few recipes that he felt were useful but missing in his grandfather's text, citing the work of Leonardo Fioravanti and Alessio Piemontese.[65] In addition to 'bring[ing] together' information from the *Garden of Gentlemen* and other texts, Vallés also 'put in order' the structure of the topics within the collection and imposed the idea that the surfaces of the body should take precedence over any other treatments.

Another important point of departure from the *Garden of Gentlemen* is revealed in de Sada's introduction. Unlike the earlier text, whose title signalled

---

62 Cabré, 'Beautiful Bodies'.

63 For a closer comparison of these two texts (including chapter concordaces), see Cabré, 'Los consejos para *hermosear*', pp. 185–202. It is well known that printed culture owed much but failed to acknowledge medieval texts; a recent study of a medieval Latin text that was a success in the early modern press, conveniently translated without recognition of the earlier source, Monica Green, 'The Sources of Eucharius Rösslin's *Rosengarden for Pregnant Women and Midwives* (1513)', *Medical History*, 53/2 (2009): pp. 167–92.

64 '… siendo la necesidad que los hombres tienen de semejantes cossas, tanta que ni en salud saben passar sin ellas ni en enfermedad pueden escusallas, havía de ser muy grande el beneficio que generalmente a todos hiziesse, quien después de havellas recogido y puesto en alguna orden las offreciesse al provecho y servicio común para que cada uno, según la calidad de su estado, pudiesse aprovecharse de la industria que en este género de policía otras personas más curiosas han alcançado'. de Sada, 'Introduction' in Vallés, *Regalo*, p. 263.

65 Vallés, *Regalo*, Book 1, Ch. 13, pp. 306–7 (Fioravanti) and Book 3, Ch. 17, pp. 391–2 (Alessio Piemontese).

a largely male intended audience, the introduction to the *Regalo* acknowledged a privileged relationship between women and beautifying practices. Pedro de Sada singled this out while summarizing the contents of the first book of his grandfather's text:

> The first book teaches very well chosen and excellent things for the service of one's own person in regards to cleaning and arraying the head, face and hands. And while it is true that it deals a little with cosmetics for women (*afeyte y compostura para mugeres*), it does that mildly so that choosing from the bad the least worst, and thus the most honest, those [women] who have this need will find remedy according to good and allowable practice.[66]

The association between women and the care of the surfaces of the body is not peculiar to the *Regalo*. In medieval Jewish, Christian and Muslim cultures, healthcare texts recognized women as knowledgeable agents of cosmetic practices.[67] However, late medieval learned practitioners such as Henri de Mondeville tended to justify their engagement with women's cosmetics by warning of the physical dangers posed by the treatments some women used: arguing that learned advice could save them from injuries that their unreliable knowledge might cause.[68] In contrast, Pedro de Sada's apology in the *Regalo* seems solely motivated by a moral concern. He constructs the authority of his grandfather's text not by appealing to his higher technical expertise but by advertising a moderate attitude to deal with the issue, in order to reconcile an uncomfortable reality – women's persistent desire for cosmetics – with a behaviour that could be widely acceptable. By doing so, he was also gendering the audience of one section of a compilation that was presented broadly as 'profitable to all'.

Early modern Iberia witnessed hot debates regarding women's adornment.[69] Particularly harsh towards women were influential intellectuals such as Juan Luis

66 'En el libro primero se enseñan cossas muy escogidas, y excelentes, para el servicio de la persona misma, en lo que toca a la limpieza y adreço de la cabeça, rostro y manos. Donde aunque es verdad que se trata algo del afeyte y compostura para mugeres, aquello es templadamente y de manera, que escogiendo de lo malo lo menor, y assí lo más honesto, podrá la que tuviere esta necessidad remediarse en ella, conforme a los buenos respetos, que pueden permitillo'. De Sada, 'Introduction' in Vallés, *Regalo*, p. 263.

67 Green, *The Trotula*, especially pp. 45–6; Cabré, 'Beautiful Bodies'; Moulinier-Brogi, 'Esthétique et soins du corps'; Carmen Caballero-Navas, 'The Care of Women's Health and Beauty: An Experience Shared by Medieval Jewish and Christian Women', *Journal of Medieval History*, 34/2 (2008), pp. 146–63.

68 McVaugh, *Rational Surgery*, pp. 215–228.

69 For the sixteenth and early seventeenth-century debates, see Pedro Tena 'La cosmética áurea a través de mujeres literarias', *Lemir*, 8 (2004). Online journal: http://parnaseo.uv.es/Lemir/Revista/Revista8/Tena2.htm [accessed: 30 May, 2010], and María-

Vives or Andrés Laguna; Vives' tough views against women's use of cosmetics were even contended by Erasmus of Rotterdam.[70] The case of Laguna is especially interesting because his overt criticism did not preclude a certain ambivalence towards the issue. When addressing the properties of lead and mercury in his Spanish annotated translation of Dioscorides, he wrote a strong condemnation on the use of cosmetics by women, particularly warning of their dangers.[71] However, throughout his work Laguna offers details of many of the cosmetic properties of the *materia medica* discussed – a trait not unrelated to his strategies to widen an audience that included women.[72]

If the control of women's appearance was a matter of concern, the regulation of the male body also took centre stage. The most sober currents of humanist thinking not only expressed contempt for women's cosmetics but also insisted that the external care (cleaning) of women's *and* men's bodies, in Juan Luis Vives' words, should be done 'without royal treats or curiosities' – precisely, the goals that the compiler of the *Regalo* and its later editor were looking to attain.[73] As the interest in beautifying recipes grew in sixteenth and early seventeenth century Iberia, anxieties over the improper care of the surfaces of the male body were expressed in literary, moral and didactic sources.[74] Castiglione's

Milagros Rivera 'Las prosistas del humanismo y del renacimiento', in Iris M. Zavala (coord.), *Breve historia feminista de la literatura española (en lengua castellana). IV. La literatura escrita por mujer. De la Edad Media al s. XVIII* (Barcelona: Anthropos, 1997), pp. 83–129, especially 119–23.

70 See Isabel Morant, *Discursos de la vida buena. Matrimonio, mujer y sexualidad en la literatura humanista* (Madrid: Cátedra, 2002), pp. 82–7. For his thinking on women's adorment, Juan Luis Vives, *De institutione feminae christianae. Liber primus.* Introduction, critical edition, translation and notes, ed. C. Fantazzi and C. Matheeussen, trans. C. Fantazzi (Leiden: Brill, 1996), Ch. 8 De Ornamentis, pp. 74–101; and *De institutione feminae christianae. Liber secundus et tertius.* Introduction, critical edition, translation and notes, ed. C. Fantazzi and C. Matheeussen, trans. C. Fantazzi (Leiden: Brill, 1996), Ch. 7 De Ornamentis, pp. 102–3.

71 Andrés Laguna, *Acerca de la materia medicinal y de los venenos* [Pedanio Dioscórides Anazarbeo] (Salamanca: M. Gast, 1566), bk. 5, Ch. 62, pp. 536–7 (lead); bk. 5, Ch. 69, pp. 542–3 (mercury). In 1616 his commentary was translated into English and published as a separate letter in Arnold Tuke's *A Discourse Against Painting*, see Snook, 'The Beautifying Part of Physik', p. 13.

72 Consolación Baranda, 'Los lectores del Dioscórides: Estrategias discursivas del Doctor Laguna', *Criticón*, 58 (1993): pp. 17–24.

73 'La limpieza del cuerpo sin regalos ni curiosidades ayuda a la salud y al ingenio: que sin falta se encoge estando suzio el cuerpo'. Juan Luis Vives, preface to *Introducción a la sabiduría*, Anvers, 1551, fol. 10r, as cited by María José Ruiz Somavilla, *El cuerpo limpio*, p. 84.

74 Mar Martínez-Góngora, 'Entre el rigor humanista y la estética cortesana: El ideal de conducta masculina en la *Respuesta de Boscán a Don Diego de Mendoza*', *Bulletin of Hispanic Studies*, 78/4 (2001): pp. 421–38 (p. 435).

count Ludovico de Canosa stated that the face of his ideal courtier should not be excessively treated and effeminated and condemned those men that use women's beautifying techniques on themselves.[75] After devoting a chapter to the great bravery, strength and masculinity of the Great Tamerlane in his *Silva de Varia Lección*, Pedro Mexía recounted the story of the Roman emperor Heliogabalus, a man who embodied not male virtue but male vice. He was 'the most effeminated and over-treated man' ever ('*más afeminado y más regalado*'). He used cosmetics and painted his face, even desiring so much being a woman that he called the greatest physicians and surgeons of his day to ask them to cut up and intervene in his body as they pleased, so that he could behave like a woman ('*usar como mujer*').[76] A learned diatribe by Brother Antonio Marqués against women's cosmetics dated in 1617 devoted a long chapter to explain why 'it is much more indecent for men than it is for women to use cosmetics and adorn themselves'; that is why men who use cosmetics 'change into women's state'.[77] Those who condemned cosmetic practices considered them women's propriety; accordingly, men interested in them were designated as 'effeminate' (that is, similar to women) a clear indication that cosmetics did not belong to them.

In this context, male beards became an important asset, for they were a powerful measure to construct proper physical representations of masculinity that maintained distinctive physical traits between women and men. As Alonso López de Corella put it in 1547: 'it is now more common to grow beards than what has been for many years'.[78] This distinction was not exempt from hierarchical value, since beards made men not just different but also 'more venerable than women'.[79]

---

[75] The Spanish translation was published in 1534, Baltasar de Castiglione, *El cortesano*, trans. Juan Boscán and intr. Rogelio Reyes Cano (Madrid: Espasa-Calpe, 1984), Ch. 4, pp. 97–8.

[76] Pedro Mexía, *Silva de varia lección*, ed. Isaías Lerner (Madrid: Castalia, 2003), Ch. 29, pp. 434–42, (pp. 434 and 437 respectively), from the last edition emended by the author in 1550–51.

[77] Ch. 8, 'Es muy más indecente afeitarse y engalanarse los hombres que las mujeres'. Fray Antonio Marqués, *Afeite y mundo mujeril*, intr. and ed. Fernando Rubio (Barcelona: Juan Flors, 1964), pp. 72–85 (p. 76): 'Sobre este desorden … añaden otro mayor los hombres, que es andar afeitados, con que pasan al estado de mujeres'.

[78] '… la naturaleza, la cual siempre es curiosa en dar la forma del cuerpo conveniente a las costumbres, dio al hombre la barba; la cual es indicio de cuánta más veneración sea el hombre que la mujer. Esto dice Galeno. Lo cual en la edad presente puede ser mejor admitido que en otra; porque es uso más común de criar barbas que muchos años ha sido'. Alonso López de Corella, *Secretos de filosofía y astrología y medicina y de las cuatro matemáticas ciencias*, intr. and ed. Juan Cruz, trans. Idoya Zorroza (Pamplona: Gobierno de Navarra, 2001), question 113, p. 258, col.a.

[79] Ibid.

The Secret Lives of Beauty Recipes

Whether explicitly or not, household recipe collections conceived for the press contained many secrets – 'things kept quiet and under cover' – that lay compilers and editors were not always ready to tell.[80] Analyzing what they presented shows that beauty was an important concern within the domestic sphere. Moreover, the attention to beauty was not marginal but widely shared, since both women and men appropriated beauty recipes for their own use in the everyday household setting. Single and clustered beauty recipes travelled from script to script, from print to print, and from print to script; but most importantly, they chart hybrid genres of writing in which these traditional categories collapse.

Beautifying recipes attest to changes in the cultural embodiment of sexual difference. Preliminary comparisons between early modern and medieval traditions seem to show a trend: a renewed interest in the sixteenth century for the care of the male body surfaces, particularly regarding the beauty of the hair of the head and beard. Comprehensive investigation of recipes for embellishing the body – who kept them, how they are grouped, the issues they encompass and the audiences they sought – shows how Iberian collections intended for print privileged certain earlier traditions over others. The *Regalo de la vida humana* aimed at a wide readership of men and women while earlier collections were more clearly gendered. While copying extensively from a male-authored source, its compiler and editor added a significant number of recipes to their text, producing a much longer and systematic collection. Nevertheless, they did not enrich the variety of beauty treatments included in the texts from which they copied. And in fact, neither the *Garden of Gentlemen* nor the *Regalo de la vida humana* contains certain beauty treatments – baths, red powders for the face, ointments for the care of the female breasts, and so on – that had a significant presence in medieval cosmetic traditions. Comparing these two texts makes also clear that the *Regalo* hid certain women's traditions that had been widely acknowledged in medieval texts.[81] Whose bodies gained or lost in the narrowing diversity of treatments on offer remains to be thoroughly studied.

Tracing back beauty recipes is not simply an exercise of erudition. It is a historical method that allows us to identify different traditions of knowledge that textual developments privilege or hide, in complex processes of communication and transformation that may journey from oral medieval communities to healthcare practitioners' textbooks and from those to learned Renaissance audiences.[82] Much work needs surely to be done. However, this preliminary attempt already

[80] Covarrubias, *Tesoro*, 2nd part, fol. 172r.

[81] Montserrat Cabré, 'Los consejos para hermosear', pp. 191–202.

[82] Peter Murray Jones has documented rich chains of communication of healthcare knowledge among the medieval laity, textually recorded when arrived at the ears of health practitioners, 'Mediating Collective Experience: The Tabula Medicine (1416–1425) as a Handbook for Medical Practice', *Between Text and Patient: The Medical Enterprise in*

shows certain themes: on the one hand, how early modern texts deserted earlier genealogies of knowledge while creating their own; on the other, how texts still bear footprints of a past that, because of those imprints, is not totally gone.

Medieval & Early Modern Europe, ed. Eliza Glaze and Brian Nance (Florence: Edizione de Galluzzo, 2011), pp. 279–307.

Chapter 9

Secrets to Healthy Living: The Revival of the Preventive Paradigm in Late Renaissance Italy

Sandra Cavallo*

The last few years have seen an extraordinary increase in the interest in recipe books, also known in Renaissance Italy as books of secrets.[1] In the Italian peninsula the first books of secrets began to be published in the 1520s, and some of them, such as the slim *Dificio di ricette* or the more substantial *Secreti del reverendo donno Alessio Piemontese* (hereforth *I secreti*), went through

* Research for this chapter was funded by the Wellcome Trust as part of the project grant 'Healthy Homes, Healthy Bodies in Renaissance and Early Modern Italy'. I am grateful to Marilyn Nicoud, Carole Rawcliffe and the editors of this volume for their valuable comments to a first version of this chapter.

1 William Eamon, *Science and the Secrets of Nature: Books of Secrets in Medieval and Early Modern Culture* (Princeton, NJ: Princeton University Press, 1994); Lynette Hunter, 'Women and Domestic Medicine: Lady Experimenters, 1570–1620', in Lynette Hunter and Sarah Hutton (eds), *Women, Science and Medicine, 1500–1700: Mothers and Sisters of the Royal Society* (Stroud: Sutton Publishing, 1997), pp. 89–107; Sara Pennell, 'Perfecting Practice? Women, Manuscript Recipes and Knowledge in Early Modern England', in Victoria E. Burke and Jonathan Gibson (eds), *Early Modern Women's Manuscript Writing* (Aldershot: Ashgate Publishing, 2004), pp. 237–55; Sara Pennell, 'Introduction', *Women and Medicine: Remedy Books 1533–1865* (Thomson Gale microfilms, 2004); Elaine Leong, 'Receipt Books, c. 1575–1800, From the Folger Shakespeare Library, Introduction'. Online: www.adam-matthew-publications.co.uk/collections_az/Receipt-Books/editorial-introduction.aspx; Elaine Leong and Sara Pennell, 'Recipe Collection and the Currency of Medical Knowledge in the Early Modern Medical Marketplace', in Mark S.R. Jenner and Patrick Wallis (eds), *Medicine and the Market in England and its Colonies, c. 1450–c. 1850* (New York, NY and Basingstoke: Palgrave Macmillan, 2007), pp. 133–52; Catherine Field, '"Many Hands Hands": Writing the Self in early Modern Women's Recipe Books', in Michelle Dowd and Julie Eckerle (eds), *Genre and Women's Life Writing in Early Modern England* (Aldershot: Ashgate Publishing, 2007), pp. 49–63; Allison Kavey, *Books of Secrets. Natural Philosophy in England 1550–1600* (Urbana, IL and Chicago, IL: University of Illinois Press, 2007); Elaine Leong, 'Making Medicines in the Early Modern Household', in *Bulletin of the History of Medicine*, 82/1 (2008): pp. 145–68; Tessa Storey, 'Italian Books of Secrets Database: Study Documentation'. Online: https://lra.le.ac.uk/handle/2381/4335.

several editions.[2] But there was also a handwritten tradition of collecting secrets that did not die out with the advent of print, and numerous exemplars of these domestic manuscripts are preserved in family archives and libraries.[3] These printed and handwritten collections of recipes contain both medical remedies for the treatment of a variety of health disorders and instructions for making a range of products for daily care of the household and the body. While recipes for preserving food are also common, in sixteenth-century Italian texts culinary recipes are notably absent.

Recent research on recipe books, carried out in particular in the English context, has focused on therapeutic remedies, employed to restore health in the sick body, and on culinary recipes – these, too, considered of medicinal value due to the close association between food and health that characterizes early modern medicine.[4] Indeed, to historians of medicine, recipe books have offered proof that the daily management of health was a domestic affair and the household a site of medical knowledge, production and experimentation. Hence these manuals have been studied for what they reveal about usage and about the routes through which medical wisdom was exchanged between patients and transferred from professional practitioners to lay people. This focus on domestic medicine has diverted attention away from the non-medical, or, rather, non-curative, components that represent a significant element of the recipe books.

Preservative remedies, which are not aimed at treating particular ailments but at keeping healthy and securing longevity, are certainly a significant component in Italian books of secrets. However, we should refrain from overstating the 'Italianness' of this focus on prevention. The phenomenal success that Piemontese's

---

[2] *Opera nuova intitolata dificio di ricette* (Venice, 1525); Alessio Piemontese, *Secreti del reverendo donno Alessio Piemontese* (Venice, 1555). The first edition of Piemontese's *Secreti* only consisted of part one of the work. Parts two and three were published in separate books in 1559, and part four appeared in 1620. The editions used here are: *La seconda parte de i secreti di diversi eccellentissimi uomini* (Milan, 1559); *Dei secreti di diversi eccellentissimi huomini ... parte terza* (Milan, 1559); *De Secreti del R.D. Alessio Piemontese parti quattro* (Venice, 1674). On these texts see Eamon, *Science*, Chapter 4; Sandra Cavallo, 'Health, Hygiene and Beauty', in Marta Ajmar-Wollheim and Flora Dennis (eds), *At Home in Renaissance Italy* (London: V&A Publishing, 2006), pp. 174–87 and 358–9; Storey, 'Italian Books of Secrets'.

[3] Only a minority of these have been published in full or in part, among them: Giuseppe Palmero (ed.), *'Et io je onsi le juncture': un manoscritto Genovese tra quattro e cinquecento: medicina, tecnica e quotidianità* (Recco: Le Mani, 1997); Andrea Aromatico and Marcella Peruzzi (eds), *Medicamenti, pozioni e incantesimi del ricettario magico urbinate* (Fano: Editrice Fortuna, 1997); Antonio P. Torresi, *Il ricettario Bardi. Cosmesi e tecnica artistica nella Firenze medicea* (Ferrara: Liberty House, 1994); 'Adunanza di Ricette', in Giuseppe. L. Passerini (ed.), *Ricette da fare bella* (Florence: Giuntina, 1912). For examples of unpublished manuscripts see those in the Wellcome Library analyzed by Storey, 'Italian Books of Secrets' and below fn. 9.

[4] Leong, 'Introduction'; Field, 'Many Many Hands'; Pennell, 'Introduction'.

text enjoyed across Europe suggests that its contents were popular well beyond the Italian peninsula.[5] Less than a third of the 350 or so recipes that make up the first edition of his *I secreti* are therapeutic; the others largely addressed the making of products relevant to the complex management of the early modern household and the safeguard of its members' health.[6] We see, for example, instructions to make powder that 'lengthens life' and another for a liqueur that 'conserves, strengthens, returns natural heat and energy, mak[ing] you vigorous of body, head and mind, giving colour to the face and sweet breath, and keep[ing] you young and robust.'[7] In Italian domestic manuscripts, too, we come across recipes for 'guaranteed health the whole year round' and remedies that secure the thorough cleansing of the various organs of the body and hence their functionality.[8] An example, a seventeenth-century Pisa manuscript presents instructions for a drink that:

> perfectly purges the kidneys, promotes urination and abundant spitting and nose cleansing, clears out the brain and cleans the lung, the liver and the spleen, expels all internal foulness, fights off headaches, gravel or newly formed stones, any quartan fever, or long-standing tertian fever, any kind of colic and pain in the sides, any kind of scabies or itch, weakness of the limbs or sluggishness, stimulates the appetite, aids sleeps, refreshes, puts weight on, fortifies, invigorates the senses and preserves health.[9]

This recipe emphasizes how efforts to preserve the existing state of health and to prevent the development of disease were inextricably connected. The regular intake of remedies like this was recommended at every change of season or for 15 days twice a year, but in some cases daily consumption was advisable. The same manuscript claims that juniper berries macerated in white wine and eau de vie and taken every day over the space of 60 years was the secret of Gregory XIII's sustained good health. In the copy of the recipe he donated to a friend in 1638, cavalier Lunadoro assured the recipient that whilst using the juniper remedy the pope never suffered from any disease and confirmed that he himself was a regular user of the remedy.[10]

Admittedly, these life-prolonging elixirs are not very numerous in books of secrets. However, if we consider the many recipes for purifying the ambient air or for perfuming bodies, clothing and furnishings, and the dozen remedies to maintain and protect skin, hair, teeth and gums that fill both the manuscript and

---

[5] This was translated into six languages, including English, within a few years from its first appearance, and then frequently reprinted. Eamon, *Science*, p. 252.

[6] Ibid., pp. 144–6.

[7] Piemontese, *Secreti*, part 1, pp. 2 and 8, and part 3, p. 5.

[8] '*E io ge onsi le juncture*', p. 35.

[9] Archivio di Stato di Pisa, Upezzinghi, Deposito Rasponi, 516, pp. 48–9 (of unfoliated manuscript).

[10] Ibid., p. 22.

printed recipe books, we find further confirmation that health preservation, and not just cure, is at the heart of the collection and dissemination of Italian 'secrets'.[11] More than has been acknowledged so far, therefore, books of secrets include secrets to healthy living. This chapter will begin by probing this understudied type of preservative 'secret'; it will highlight the links between the books of secrets and another printed genre in the vernacular that enjoyed much popularity in the sixteenth and seventeenth centuries: the regimen of health. Regimens provided comprehensive recommendations about the healthy habits to be adopted in various spheres of life in order to prevent disease. Although the two genres remained formally separate both in terms of authorship and contents (regimens consist of advice about healthy lifestyles while books of secrets are collections of recipes), they share to some extent similar concerns. They both participate in the culture of prevention which, as I shall argue, pervades so many practices and forms of cultural expression in this period.

Preventive Recipes

The widely popular regimens of health, the *consilia* of individual doctors, and the occasional illustration of practice in recipe books all clarify the prophylactic role of many of the remedies and provide the rationale for understanding their use within early modern healthy living. The numerous recipes for the fabrication of scented pastes, waters, powders and oils certainly have a preventive function. These were regularly employed to perfume the body (especially the exposed parts such as the face and the hands), the clothes and bedding in which the body rested, and all sorts of household linen including tablecloths and napkins.[12] Perfumed pastes were also used to make necklaces, earrings, buttons and even paternosters (rosary beads) that emanated pleasant smells while in use. Little bags filled with perfumed powder were employed to sweeten linen and clothes stored in chests. The ambient air within a house was also sweetened by sprinkling or burning domestically produced scented substances in fires, perfume-burners, lanterns or 'incense bricks for a room'.[13]

These procedures were all designed to protect individuals from catching diseases that were supposed to travel through the air and be absorbed into the body through breathing, the activity that one could not dispense with. Infections, especially epidemics, were traditionally thought to be caused by stinking vapours

---

[11] The following analysis is based on information contained in the Italian books of secrets database. Online: https://lra.le.ac.uk/handle/2381/4335.

[12] For evidence of the habit of perfuming table linen see Domenico Romoli, *La singolare dottrina di M. Domenico Romoli ... nel fine un breue et notabile trattato del reggimento della sanità* (Venice, 1560), pp. 9 and 10. Romoli also recommends to place the perfume burner on the table covered with food.

[13] Piemontese, *Secreti*, part 1, pp. 90 and 92.

rising from rotten matter or putrid water, and in a period in which the plague was a recurrent phenomenon, these ancient theories led to a new concern for the effects of air on health.[14] The air one breathed was seen as the main vehicle of infection. Some, following Plato, believed that the bad vapours were not immaterial, but a real substance whose particles could affect people even at a distance from the source that had generated them. Others, following Aristotle, believed that though foul odours were immaterial their harmful qualities altered the nature of the air and could therefore be carried a long way.[15] Aromas were therefore used extensively to create a safe environment in the home through the perfuming of rooms, while the fashion for perfumed clothes and accessories to be worn aimed to envelope the body in a thick layer of fragrances that travelled at all times with the wearer and could be breathed continuously to combat the attack of the noxious air.

In both cases, the focus was on breathing, and although some have suggested that the tainted, disease-carrying air was believed to penetrate the body through the pores, or even through its various orifices, it should be noted that the emphasis in medical theory and advice is disproportionately on the need to rectify the air that is inhaled through the nostrils.[16] Why? The *consilia* of fifteenth-century Italian doctors systematically noted the vulnerability of the brain in relation to the dangers of unhealthy air.[17] Breathing in purified air was crucial precisely because it was believed that the brain, not the nose, was the olfactory organ (the nose was just a conduit that conveyed the air to the brain). Hence the quality of the air inevitably affected one of the three cardinal organs of the body; thus it affected the delicate sensory and psychic faculties, as well as the function of movement, for which the brain was responsible through the spinal cord and the nerves.[18]

---

[14] On the frequency and localization of outbreaks of plague in the Italian peninsula, see Lorenzo Del Panta, *Le epidemie nella storia demografica Italiana (sec. XIV–XIX)* (Turin: Loescher, 1980), p. 117 ff.

[15] Simon Kemp, 'A Medieval Controversy about Odour', *Journal of the History of the Behavioral Sciences*, 33/3 (1997): pp. 211–19.

[16] For an excessive emphasis on penetration of disease through the skin, see Georges Vigarello, *Concepts of Cleanliness: Changing Attitudes in France since the Middle Ages* (Cambridge: Cambridge University Press, 1988), Ch. 1. As recently argued by Palmer, the air remained the principal preoccupation in spite of the new importance attributed to contagion. Richard Palmer, 'Gerolamo Mercuriale and the plague in Venice', in Alessandro Arcangeli and Vivian Nutton (eds), *Girolamo Mercuriale. Medicina e cultura nell'Europa del* '500 (Florence: L.S. Olschki, 2008), pp. 51–65, especially p. 64.

[17] Richard Palmer, 'In Bad Odour: Smell and its Significance in Medicine from Antiquity to the Seventeenth Century', in William F. Bynum and Roy Porter (eds), *Medicine and the Five Senses* (Cambridge: Cambridge University Press, 2004), pp. 51–68 (pp. 63–4).

[18] On the form taken by this ternary physiology in the Middle Ages and on the role attributed to the brain in this system see Danielle Jacquart, 'Coeur ou cerveau? Les hesitations medievales sur l'origine de la sensation et le choix de Turisanus', *Micrologus*, 11 (2003): pp. 73–96.

The need to protect the brain and the neurological functions over which it presided was also at the heart of those recipes aiming to improve the breath – from sweet meats (*moscardini*) to be held in the mouth to scented mouthwashes. The bad breath released from the mouth, clearly a sign that something was rotten inside the body, represented a health hazard, since there was a risk of damaging the brain and nervous system merely by inhaling the noxious vapours exhaled from the mouth. Presumably, similar fears lay behind the many deodorant remedies circulating in the period; as well as being unpleasant, the bad odour released by 'those who stink when they sweat' or those who have stinking feet and armpits was perceived as a threat to health.[19]

Indeed, preoccupations with decorum and health concerns are closely allied in many of the recipes. Dental care was a crucial theme in these texts. The dozens of instructions for making toothpastes and little sticks for cleaning the teeth, have both cosmetic and preventive functions. Dental remedies served dual roles – they both whitened teeth to make them beautiful and kept them healthy and firm by preventing bad breath and preserving the gums.[20] Likewise the plethora of recipes for skin and hair care also had both cosmetic and prophylactic aims, and we can see this combination of purposes in the remedies designed to prevent the sun 'from damaging the face', or 'from harming you on the head when you travel'.[21] The same can be said of cleaning agents employed in routine hygienic procedures. A 'lessia' to wash the hair 'comforts the brain and memory, as well as making the hair grow and go blond', and a soft paste for washing hands is good for the chapping of the hands, the mouth and the nose, for keeping the skin soft and scented but it also 'comforts the memory'.[22]

Medicine as a Preventive Art

Along with remedies designed to heal the sick body, prophylactic recipes and advice do therefore occupy considerable space in the literature of secrets. The medical significance of hygiene and beauty-care operations is made clear in the short regimen of health included in the third volume of Alessio's *Secreti*. The text prescribes, for those aiming to keep healthy, different rules of conduct for each month of the year: alongside detailed recommendations concerning diet, the use of blood-letting, sex and exercise, these also advise against washing the head except in January and October, they suggest to wash the face often in May, while going to the baths should take place in March but be avoided in

[19] Piemontese, *Secreti*, part 4, p. 417 and part 2, p. 56.

[20] Piemontese, *Secreti*, part 1, pp. 125 and 127. Some recipes also specify frequency of usage, for example 'do this every eight days at least', p. 126.

[21] *Centuria di secreti politici, cimichi e naturali* (Venice, 1626), p. 52; Piemontese, *Secreti*, part 2, p. 27.

[22] *Centuria.*, p. 118; Piemontese, *Secreti*, part 2, p. 45.

November.[23] The preventative value of hygienic practices was also advocated by contemporary doctors in another genre of medical text – the regimen of health. For example, the morning toilette was presented in regimens of health as a key procedure enabling the various organs and the scalp and the skin of the upper body to be cleansed of the waste produced daily during the ternary digestive process that converted food into blood, tissues and other corporeal matters. If these superfluities were to accumulate in the body, and especially in the vulnerable head, they could produce obstructions and, by rotting inside, degenerate into morbid vapours and humours.

In classical medicine and medieval commentaries, the foundation of Renaissance medical thought and practice, prevention fulfilled a role equally important to that of treatment. Avicenna's *Canon*, the most popular medical textbook in the Renaissance, held that medicine and what the ancients called 'hygiene', that is the art of taking care of one's health, were two complementary practices.[24] In Hippocratic treatises, as well as in Galen's works, medicine was not just conceptualized as a healing art but as the art of well-being, of conserving health. While Hippocratic medicine gave a key role to the balance between nourishment and exercise, Galen significantly expanded the factors to be attended to for the sake of preserving health. For Galen, moreover, the task of the physician was not limited to treating disease, but included indications for the correct way of living as well as dietetic methods (where 'diet' stands for the ideal lifestyle) that would enable to stay in good health.[25] These ideas were then systematized in the Middle Ages. In particular, a dietetic doctrine began to be established in the Arabic texts that became available to Europeans through the translations undertaken in Italy and Spain in the eleventh and twelfth centuries and was further consolidated in the regimens of health compiled in Europe in the thirteenth and fourteenth centuries by figures such as Petrus Hispanus, Taddeo Alderotti, Aldobrandino of

[23] The calendar form of this short regimen draws on the tradition of purely practical texts established by the pseudo-Hippocratic *Du regime*. The genre was common in the High Middle Ages but died out subsequently when the translation of Arabic and Hispanic regimens established new models for the literature of health advice. Marilyn Nicoud, *Les régimes de santé au moyen age*, I–II (2 vols., Rome, 2007), vol. 1, pp. 15–6.

[24] Pedro Gil-Sotres, 'The Regimens of Health', in Mirko D. Grmek (ed.), *Western Medical Thought from Antiquity to the Middle Ages* (Cambridge, MA: Harvard University Press, 1998), pp. 291–396.

[25] Nicoud, *Les régimes*, pp. 2–6; Carole Rawcliffe, 'The Concept of Health in Late Medieval Society', in Simonetta Cavaciocchi (ed.), *Le interazioni fra economia e ambiente biologico nell'Europa preindustriale. Secc. XIII–XVIII* (Florence: Florence University Press, 2010), pp. 321–38 (p. 322). On Galen, see Klaus Bergdoldt, *Wellbeing. A Cultural History of Healthy Living* (Cambridge: Cambridge University Press, 2008), pp. 87–93; Heikki Mikkeli, *Hygiene in the Early Modern Medical Tradition* (Helsinki, Finnish Academy of Science and Letters, 1999), pp. 29–38; and Galen, *De sanitate tuenda*, in *A Translation of Galen's Hygiene*, trans. R. M. Green (Oxford: Blackwell Publishing, 1951), Ch. 1.

Siena, Arnaldus de Villa Nova and Maino de Maineri.[26] Some authors, such as Aldobrandino of Siena, went as far as to suggest that the art of healthy living was the main constituent of the science of physics.[27] At the onset of the Renaissance prevention was, therefore, seen as a branch of medicine, and the idea was further theorized by sixteenth-century academic medicine.[28]

Though the preventive paradigm appears firmly established in medical theory at the beginning of the early modern era, with few exceptions medical historians of the period have concentrated on curative medicine, healing practices (professional, empiric and domestic) and medicinal remedies while largely disregarding the measures actively designed to preserve health.[29] This tendency to focus on care of the sick body reflects perhaps the enduring concern of modern western medicine for cures rather than prevention. However, health-preserving measures have also been marginalized as a consequence of historians' emphasis on the 'medical marketplace', especially in the highly commercialized and consumer-oriented economy of Renaissance Italy.[30] Recent commentators have, in fact, extended back to the Renaissance the rise of a medical consumerism initially associated with the eighteenth century.[31] They argue that the Italian medical economy boomed as early as the sixteenth century, accelerating the demise of traditional behavioural regimens, which stressed the importance of a healthy lifestyle to maintain and restore health. In the Italian peninsula this paradigm supposedly gave way to a new attitude to health, whereby cures available on the market and the possibility of eliminating disease through treatment rather than the need to keep the body 'temperate' became the main focal points.[32]

---

26 On this process, see Gil-Sotres, 'Regimen' and Nicoud, *Les régimes*.

27 Nicoud, *Les régimes*, p. 117.

28 Mikkeli, *Hygiene*, pp. 45–56.

29 Notable exceptions are Andrew Wear, *Knowledge and Practice in English Medicine, 1550–1680* (Cambridge: Cambridge University Press 2000), Ch. 4; Carole Rawcliffe, '"Delectable Sightes and Fragrant Smells": Gardens and Health in Late Medieval and Early Modern England', *Garden History* 36/1 (2008): pp. 3–21. See also note 36 below.

30 On urban prosperity and consumer behaviour in Renaissance Italy, see Richard A. Goldthwaite, *Wealth and the Demand for Art in Italy, 1300–1600* (Baltimore, MD: The Johns Hopkins University Press, 1993) and Lisa Jardine, *Worldly Goods: A New History of the Renaissance* (Basingstoke: Nan A. Talese, 1996).

31 On the historiography on the medical marketplace and its limitations see Mark Jenner and Patrick Wallis, 'Introduction', in Jenner and Wallis, *Medicine and the Market*, pp. 1–23.

32 This view is summarized in Eamon, 'Markets, Piazzas and Villages', in Lorraine Daston and Katharine Park (eds), *Cambridge History of Science*, vol. 3: *Early Modern Science* (Cambridge: Cambridge University Press, 2006), pp. 206–23 (p. 211) and Eamon, 'The Scientific Renaissance', in Guido Ruggiero (ed.), *A Companion to the Worlds of the Renaissance* (Oxford: Blackwell Publishing, 2002), pp. 403–24 and 412–14. See also Eamon, *Science*, pp. 412 and 414.

Various factors contributed to this shift: the arrival from the East and later from the New World of hitherto unknown exotic substances that were commercialized on the market as repositories of wondrous medicinal properties; the influence of Paracelsus' theories, defining disease as an entity rather than the result of a humoural imbalance; the proliferation and appeal of charlatans and empirics who sold ready-made remedies designed to treat the impersonal disease rather than the individual body and who often used the rhetoric of secrecy to promote their nostrums.[33] Taken together, these elements seem to lead to a new confidence in the possibility of targeting specific illnesses, which replaced and invalidated the idea that health depended on diet, hygiene and the maintenance of a good humoural balance. For why give so much value to a healthy lifestyle if disease could be treated?

More or less explicitly, therefore, the preventive paradigm has been viewed as already in decline in this period, and not just at the popular level. It has also been suggested that prevention was disregarded by sixteenth-century Italian and French physicians, who considered it peripheral to medicine and no longer complementary to therapy.[34] It would seem that we are already in a modern ontological framework, in which the sick body rather than the healthy (or imperfectly healthy) body is the main object of medicine.

However, in reality many elements hint at the persistence of the preventive paradigm and its pervasiveness. The proximity between medicinal and health-preserving products typical of the books of secrets can be found in more commercial spheres of medical practice, such as charlatanism. Though characterized as sellers of medical nostrums, charlatans also included in their wares an array of goods for hygiene, cleanliness and body care. Women in particular, a regular presence in mountebank troupes, dispensed advice about the care of skin, hair and teeth, especially to the female clientele.[35] But the lasting nature of preventive concerns is also evident in the success that health advice literature enjoyed after the introduction of print. In Italy the genre had its heyday in the guides to healthy living published both in Latin and, more frequently, in the vernacular during the

---

33 On the expansion of pharmacists and on the influence of Paracelsus in Italy see Richard Palmer, 'Pharmacy in the republic of Venice in the sixteenth century', in Andrew Wear, Richard K. French and Iain M. Lonie (eds), *The Medical Renaissance of the Sixteenth Century* (Cambridge: Cambridge University Press, 1985), pp. 100–118; on charlatans see David Gentilcore, *Medical Charlatanism in Early Modern Italy* (Oxford: Oxford University Press, 2006).

34 Nancy Siraisi, *The Clock and the Mirror: Girolamo Cardano and Renaissance Medicine* (Princeton, NJ: Princeton University Press, 1997), p. 73; Jean Céard, 'La diététique dans la medicine de la renaissance', in Jean-Claude Margolin and Robert Sauzet (eds), *Pratiques et discours alimentaires à la renaissance* (Paris: G.-P. Maisonneuve et Larose, 1982), pp. 21–36 and especially pp. 22–4.

35 M.A. Katritzky, *Women, Medicine and Theatre, 1500–1750: Literary Mountebanks and Performing Quacks* (Aldershot: Ashgate Publishing, 2007), pp. 122 and 166.

sixteenth and seventeenth centuries.[36] The latter addressed a larger audience and the remainder of this chapter will concentrate on these popular books.

Vernacular Guides to Healthy Living: Trends, Authors and Editorial Strategies

While individual regimens in the vernacular were already appearing in print at the turn of the sixteenth century, the number of new titles and editions increased substantially from the 1540s onwards and only began to decline in the last quarter of the seventeenth century, when the genre assumed different forms. For a century and a half, these handbooks of healthy living exercised considerable appeal for readers. The timing of their rise and the popularity they enjoyed thus parallel those of books of secrets discussed earlier in this chapter.

Yet regimens differ substantially from the books of secrets in many respects: for one thing they are mostly authored by physicians, while the most common compilers of the books of secrets were jesters, charlatans and professional writers (*poligrafi*).[37] Moreover, though both offer examples of health-related advice, the kind of knowledge they popularize is different. Books of secrets claim to uncover the secret powers of nature by revealing the occult properties of miraculous substances or combinations of substances.[38] Vernacular regimens, on the other hand, do not unveil esoteric wisdom but disseminate time-honoured medical knowledge previously only accessible to the learned and the powerful and promote the transformation of this knowledge into everyday practices of healthy living, to be followed by ordinary readers. This different agenda is perhaps the reason why the word 'secret' is no longer used in the Renaissance in relation to regimens, though one could say that such a connection had authoritatively been established by the *Secretum secretorum*, an extremely popular text in the medieval Latin West, which included among its varied 'secrets' a rule of health. In sixteenth-century Italian printed and manuscript collections of recipes, however, the word 'secret' is commonly employed as a synonym of 'recipe'. Contemporary regimens only occasionally include a recipe, the advice they contain being of a different

36 For a Europe-wide overview on regimens: Harold J. Cook, 'Physical Methods', in William F. Bynum and Roy Porter (eds), *Companion Encyclopedia of the History of Medicine* (2 vols; London: Routledge, 1993), vol. 2, pp. 939–60; and Andrew Wear, 'The History of Personal Hygiene', in ibid., pp. 1283–308; Virginia Smith, *Clean: A History of Personal Hygiene and Purity* (Oxford: Oxford University Press, 2007). On dietary advice, see Ken Albala, *Eating Right in the Renaissance* (Berkeley: University of California Press, 2002); Rudolph Bell, *How to Do It: Guides to Good Living for Renaissance Italians* (Chicago, IL: University of Chicago Press, 1999). On the genre in Europe see note 36 and Alessandro Arcangeli, *Recreation in the Renaissance: Attitudes towards Leisure and Past Times in European Culture 1425–1675* (Basingstoke: Palgrave Macmillan, 2003).

37 See Gentilcore, *Medical Charlatanism*, pp. 360–62.

38 See Eamon, *Science*, pp. 196–233.

nature, referring to practices of healthy living rather than to the manufacture and application of a remedy.

The bulk of these texts consist of detailed recommendations about the precautions to be taken in the daily management of the body and in the running of the household to promote the health of its members. They prescribe when and how one should sleep, purge and exercise, what and when one should eat and drink, when a man should have sexual intercourse, and which emotions are beneficial or harmful to health. Some of the titles also include recommendations about the domestic setting and discuss the ideal location of the home in terms of winds, temperature, humidity, and the way in which it should be built, furnished and managed to protect the inhabitants from breathing harmful air. Clearly these books are heirs to the tradition of ancient Arabic and medieval treatises devoted to the regulation of the six 'necessary' or 'non-natural' things on which health depends: the air one breathes, the food and drink one ingests, sleeping and waking, movement and rest, evacuation and repletion, and the passions of the soul.[39]

Some of the early modern printed regimens are indeed vernacular translations of ancient or medieval texts. Among these we find in 1538, under the title *Col nome di dio*, the translation of the *Secretum secretorum*, a composite pseudo-Aristotelian epistolary text that had been available in Latin in manuscript form since the thirteenth century, and which also contained a regimen of health for the prince.[40] This was followed, in 1549, by the first translation into a vernacular language of Galen's *De sanitate tuenda* (*Di Galeno*) and, in the same year, by the vernacular rendition of a regimen erroneously presented as authored by Arnald of Villanova, in reality to be attributed to Maino de Maineri.[41] In 1550, the translation of Arnald's *De conservanda iuventute* by the physician Panfilo Fiorimbeni from Fossombrone was annexed to the Italian edition of Pictorius's regimen.[42]

---

39 On the development of the theory of the non-naturals see Lelland J. Rather, 'The "Six Things Non-Natural": Origins and Fate of a Doctrine and a Phrase', *Clio Medica*, 3 (1968): pp. 337–47; Gil-Sotres, 'Regimen'; Lluís Garcia-Ballester, 'On the Origin of the "Six Non-Natural Things" in Galen', in *Galen and Galenism* (Aldershot: Ashgate Publishing, 2002), pp. 105–15.

40 On the fortunes of the *Secretum secretorum*, see Eamon, *Science*, pp. 45–53. The text promised to reveal the great philosopher's esoteric doctrine about the art of government and moral philosophy but also contained sections on various other branches of knowledge.

41 *Opera utilissima di Arnaldo da Villanova di conservare la sanità* (Venice, 1549). For the identification of Maino de Maineri (d. 1368) as the real author see Sebastià Giralt i Soler, *Arnau de Vilanova en la imprenta renaixentista, segle XVI* (Manresa: PAHCS, 2002), pp. 18 and 163 note 9.

42 Giorgio Pictorius, *Dialogi del eccellente medico m. Giorgio Pittore Villingano, del modo del conseruare la sanita ... aggiuntoui un trattato di Arnaldo di Villa nuoua* (Venice, 1550).

This stream of learned translations, clustering in the years 1538 to 1550, was followed, in 1581, by the appearance in the vernacular of a more popular text, the puzzling *Regimen sanitatis salernitanum*, which presents, without any logical order, short pieces of advice on different aspects of healthy living in short rhymes and proverbs that could be easily memorized.[43] The didactic intent of the translation, which was intended for all those who can read, including women, is explicitly declared at the outset by the translator, Father Serafino Razzi, from the religious order of the Predicatori.[44]

A much larger proportion of the regimens published in Italian in the sixteenth and seventeenth centuries is taken up however by 'new' texts, authored by contemporary physicians. On the one hand there are the vernacular versions of works first published in Latin, such as Ficino's *Tre vite*, and the regimens authored by Pictorius, Lemnius, Petronio, Durante and Fonseca.[45] The translation (normally also by a physician) would sometimes follow the Latin edition by just a year or so (as in the case of Fonseca, Pictorius, Lemnius) and would be promoted by the same author; in other cases (Ficino, Petronio) it appeared posthumously. On the other hand, we have the many texts that were compiled and printed directly in the vernacular, the earliest example probably being Savonarola's *Libreto*, published in 1508.[46]

Interestingly, these new texts enjoyed the greatest editorial success. The most frequently reprinted regimens were in fact newly authored rather than translations of ancient and medieval authoritative texts and in the vernacular rather than Latin. The most popular was undoubtedly Castore Durante's *Il tesoro di sanità*, which first appeared in Latin in 1565 and went through at least 29 editions in Italian between 1586 and 1691, with an English translation published in 1686. This was followed, in the seventeenth century, by Auda's

---

43 Several versions of this text, erroneously attributed to the Salernitan School, exist both in the manuscript and edited traditions, to the extent that, according to Nicoud, one should talk of 'regimina sanitatis salernitanum', in the plural. Nicoud, *Les régimes*, p. 368; idem, 'Il regimen sanitatis Salernitanum. Premessa ad un'edizione critica', in Danielle Jacquart and Agostino Paravicini-Bagliani (eds), *La scuola medica salernitana, gli autori e i testi* (Florence: SISMEL edizioni del Galluzzo, 2007), pp. 365–84.

44 *Scuola Salernitana del modo di conseruarsi in sanità* (Perugia, 1587), 'Prefazione'. The involvement of an ecclesiastic in the publication of regimens of health is exceptional in the Italian context, while it seems to have been much more common in England: Rawcliffe, 'Concept of Health', p. 333.

45 Marsilio Ficino, *De le tre vite* (Latin edn 1501; Venice, 1548); Pictorius, *Dialogi* (Latin edn 1549; Venice, 1550); Levinius Lemnius, *Della complessione del corpo humano libri due* (Latin edn 1561; Venice, 1564); Alessandro Petronio, *Del viver delli Romani et di conservar la sanità* (Latin edn 1585; Rome, 1592); Castore Durante, *Il tesoro della sanità* (Latin edn 1565; Venice, 1586); Rodrigo Fonseca, *Del conservar la sanità* (Latin edn 1602; Florence, 1603).

46 Michele Savonarola, *Libreto ... de tutte le cose che se manzano ... e de sei cose non naturale & le regule per conseruare la sanita de li corpi humani* (Venice, 1508).

Breve compendio, with 17 editions in just 60 years, while many other texts went through four or five editions.[47]

It is remarkable, therefore, that the texts available in the early age of print to readers willing to learn about the measures that fostered a healthy life were largely the work of contemporary authors. Rather than promoting translations of previous works, modern doctors drew liberally on the treatises available in Latin or in manuscript form, to offer their readers personal syntheses of consolidated knowledge about healthy living. The heavy borrowing that has been documented in relation to one of the first regimens printed in the vernacular, Ugo Benzi's *Tractato vtilissimo circa lo regimento e conseruatione de la sanitade* (1481), could well apply also to subsequent examples of this genre. According to J. Hill Cotton this publication was simply the translation into the vernacular of a regimen in Latin published just six years earlier under the name of Benedetto Reguardati and presented by Benzi's publisher as the latter's work.[48] On the other hand, Reguardati himself had substantially tapped into the work of a fourteenth-century doctor, Barnabas de Reatinis.[49] There is no doubt that further philological comparison of the kind carried out by Cotton and Marilyn Nicoud would highlight many common and often identical elements in older and new regimens.[50] Although they vary from one another in structure, in the order in which the non-naturals are addressed and, above all, in the space and level of detail devoted to each of them, sixteenth-century regimens are patchworks incorporating substantial elements from earlier works.[51] But the habit of reworking existing texts and presenting them as new also concerned the most recent regimens. Re-editions with considerable additions and variations of contemporary works, once the original author was deceased, are extremely frequent. Boldo, for example, reissues Savonarola's *Libreto* in 1575 'set out in better order, expanded and amended and almost made anew'; Benzi's

---

47 Durante, *Tesoro della sanità* and Domenico Auda, *Breve compendio ... con un trattato per conservarsi in sanità* (Rome, 1652).

48 Ugo Benzi, *Tractato utilissimo circa lo regimento e conseruatione de la sanitade* (Milan, 1481). According to J. Hill Cotton this is a translation of *Pulcherimum et utilissimum opus ad sanitatis conservationem*, a text attributed to Benedetto Reguardati of Nursia (1398–1469), first published in Rome in 1475. Cotton, 'Benedetto Reguardati: author of Ugo Benzi's Tractato de la conservatione de la sanitade', *Medical History*, 12/2 (1968): pp. 175–89.

49 The *Libellus de conservanda sanitate* by Barnabas de Reatinis of Reggio was finished in Mantua in 1331. Barnabas' work supplied the source and framework for Reguardati's five introductory chapters and his final chapter. The source of the ninety chapters in between, on diet, is another treatise by Barnabas, the *Compendium de naturis et proprietatibus alimentorum* (1338). Cotton, 'Benedetto Reguardati', p. 79.

50 Nicoud, *Les régimes*, p. 77.

51 Nicoud, *Les régimes*, p. 141. On the variations in structure in medieval regimens, see also Melitta W. Adamson, *Medieval Dietetics. Food and Drink in Regimen Sanitatis Literature from 800 to 1400* (Frankfurt: P. Lang, 1995).

regimen underwent a further manipulation in 1618 when it was reissued with supplements by Lodovico Bertaudi (d. 1625), court physician of the dukes of Savoy.[52] Many other cases could be mentioned. Translators, too, were not extraneous to the process of marking the text with personal additions and were keen to signal their input. In the preface to his translation of the *Scuola salernitana* Serafino Razzi proudly alerted the reader to the new order he had imposed to the work by collating information on the same topic, organizing the entries in alphabetical order and finally trimming it of unnecessary detail and adding some words and sentences 'commonly approved by the *volgo*' and by the 'specialists in the art and profession'.[53] Even the translation of a centuries-old text could therefore constitute a relatively new work.

In one way or the other, whether as an author, a translator or an editor, many members of the medical professions were eager to associate their names with printed preventive advice addressed to non-specialists. Indeed, Italian Renaissance regimens were (unlike their English counterpart), overwhelmingly authored by learned physicians of some renown.[54] As already suggested, this is one of the elements that distance them from the books of secrets. Overall the authors seem to have been distinguished physicians and scholars, who cultivated broad humanistic interests in philosophy, poetry, botany and sometimes astrology. Many, like Savonarola, Boldo, Pisanelli, Fonseca, Durante and Panaroli held university chairs as professors of botany, anatomy or medicine at university, while others (Reguardati, Durante, Rangoni, Savonarola, Petronio) at some point of their career were employed at the courts of various Italian princes, including the pope, as physician of the household or personal attendant to the ruler. For these figures regimens were often just one part of their numerous writings on medical, moral, historical and literary topics, and, particularly in the sixteenth century, were often the only work they published in the vernacular. Certain authors such as Paschetti, Traffichetti and Viviani seem to have been less distinguished figures but were nevertheless renowned practitioners who numbered aristocrats and religious houses among their clients. In any case it would appear that authoring, editing or

---

52 Bartolomeo Boldo, *Libro della natura et nirtu delle cose, che nutriscono* (Venice, 1575); *Regole della sanità et della natura de cibi di Ugo Benzo. Con le annotazioni di Giò Ludovico Bertaudo* (Turin, 1618).

53 *Scuola salernitana*, 'Prefazione'.

54 A rare exception is represented by Auda's *Breve compendio*, put together by a Reformed Franciscan, an apothecary at the Hospital of Santo Spirito in Rome. Another anomalous case is Cornaro's *Trattato de la vita sobria* (1558), expanded as *Discorsi della vita sobria* (1591). On English regimens, see Paul Slack, 'Mirrors of Health and Treasures of Poor Men: the Uses of the Vernacular Medical Literature of Tudor England', in Charles Webster (ed.), *Health, Medicine and Mortality in the Sixteenth Century* (Cambridge: Cambridge University Press, 1979), pp. 237–73, especially pp. 252 and 255. According to Slack 'only one third of regimens came from established physicians'.

translating a regimen into the vernacular was viewed by physicians of different backgrounds as a means to promote their professional image.

What does all this tell us? It certainly complicates the opposition we tend to set up between 'learned' and 'popular' authors and the association of the former with Latin texts and the latter with a vernacular production.[55] It challenges the conventional image of learned physicians, showing that they differentiated their work, writing accessible texts for a popular audience as well as academic works for a more educated or professional one. It also testifies to the persistence among the medical profession of an idea of medicine as a science aiming to preserve health, not just to treat disease, and of an adherence to the Galenic idea of the doctor as a figure whose duties include the popularization of the principles of healthy living.

The didactic intent of Renaissance regimens is evident both in the choice of language and in the format of these books. The majority of these texts were written in the vernacular (or made available in the vernacular in the wake of a Latin edition) and issued in small, portable formats (in octavo or even duodecimo). Hence they circulated at a more affordable price than their Latin counterparts, which were often published in quarto or folio. There are instances of a large-sized first edition, possibly published to test the market and establish the prestige of the work, was soon followed by a portable one.[56] Consequently, there was a clear strategy to target a wide public of non-specialists.

These editorial strategies betray a commercial aim. The authors, translators or editors who were contributing to the multiplication of regimen titles were aware that a market existed for this kind of book; they were responding to the demand of potential consumers who did not read Latin but nonetheless evinced an intellectual thirst for greater knowledge of medical prevention than what was available in the popular *Scuola salernitana*.[57] Though the display of erudition cannot be alien to these distinguished doctors' engagement with the tradition of dietetic and hygienic advice, this was no longer a purely antiquarian topic but a subject that appealed to current concerns. While in the late fifteenth century the court milieu was the environment in which the first printed regimens in vernacular made their appearance, by the mid-sixteenth century, as the varied profile of their authors also suggests, these texts were aimed at a broader educated audience that transcended the boundaries of the court.[58]

---

55 See the discussion in Gentilcore, *Medical Charlatanism*, pp. 361–2.

56 This is the case, for example, of the texts by Savonarola, Pisanelli and Traffichetti, Bertaudo's version of Benzi's.

57 *Scuola salernitana*, the first translation into vernacular of the *Regimen sanitatis salernitanum* appeared relatively late, in 1587 (see note 44) and presented the Latin verses in blank, unrhymed verses; a new translation from Latin, this time with rhymed verses, was then published in 1630. *Scola salernitana per acquistare, e custodire la sanita* (Venezia, 1630). The text was reprinted four times up to 1712, collated with other regimens by Cornaro and Pictorio.

58 On fifteenth-century printed vernacular regimens see Chiara Crisciani, 'Histories, Stories, *Exempla*, and Anecdotes: Michele Savonarola from Latin to vernacular', in Gianna

A Culture of Prevention

This brief exploration into the printing of vernacular regimens shows that there was a sustained production of preventive discourse over the sixteenth and seventeenth centuries, not just through the publication of new titles but through new editions of established works and translations from Latin of existing texts. While there is considerable overlap between these texts, each also included significant new elements and can be seen as distinct works in their own right. The flourishing of health advice literature testifies to the survival and even perhaps to an expansion of the preventive concerns in this period. Indeed, the vitality of the culture of prevention is visible also in other, less strictly medical forms of systematization and dissemination of health knowledge.

Regimens cannot be entirely demarcated as a genre because some of the themes they discuss are also dealt with in other types of contemporary literature. This is the case for the *libri dello scalco* or *del trinciante* genre of treatises on food and table service that flourished in Italy in the sixteenth and seventeenth centuries.[59] Compiled by stewards and banquet managers in charge of overseeing the table of their lord at the court of popes, cardinals and princes, these books not only discuss the effects of particular foods and how they were cooked and eaten on health and on specific complexions but also contain sections on the other non-naturals. For example, Francesco Colle's *Refugio over ammonitorio del gentiluomo*, authored by the *scalco* of Duke Alfonso of Ferrara in 1533 and dedicated to the duke himself, includes recommendations about exercise and sleeping as well as diet. The subject of exercise, together with that of baths and frictions, is developed in complete chapters and in much greater detail in some editions of the 1560 *La singolare dottrina* by Domenico Romoli, *scalco* of Cardinal Ridolfi and then of Pope Paul III. To reinforce further the proximity between the two genres, we can note that Romoli's treatise is published with an annexed regimen of health by Roberto Gropretio.[60] Moreover, books of recommendations for the upbringing of the gentleman and the education of the young scholar, often authored by clerics, sometime include sections on how to maintain health. For example, Father Bartolomeo Meduna's *Lo scolare* (1588) contains a short treatise on the six non-naturals for the use of the young pupil.[61]

Preventive elements were also at the heart of explorations of nature in herbaria, bestiaria and lapidaria, and these centuries-old literary genres underwent a revival

Pomata and Nancy Siriaisi (eds), *Historia: Empiricism and Erudition in Early Modern Europe* (Cambridge, MA: MIT Press, 2005), pp. 297–324.

59 Gigliola Fragnito, 'La trattatistica cinque e seicentesca sulla corte cardinalizia', *Annali dell'Istituto Storico Italo-Germanico in Trento*, 18 (1991): pp. 135–85 (pp. 139–40).

60 See note 12.

61 In the book, organized in form of a dialogue, the task of advising pupils on these matters is entrusted to a physician. Bartolameo Meduna, *Lo Scolare del r.p.m. Bartolameo Meduna Conventuale* (Venezia, 1588).

during the Renaissance. In these texts, plants, stones and minerals and animal substances were described and classified according to their occult qualities and the protective as well as healing powers they possessed. Rooted in classical writings, beliefs in the therapeutic and prophylactic properties of natural objects acquired new importance by the late fifteenth-century, coinciding with the remarkable rebirth enjoyed by magic in this period.[62] Their success was also related to the use of natural objects, such as specific stones and animal substances in amulets to prevent a person falling prey of disease or to encourage conception and protect in childbirth. For example, Jacqueline Musacchio has demonstrated the increased use of talismanic natural objects in reproductive rituals in Renaissance Italy.[63] Though often categorized as 'popular', these beliefs had the full backing of the learned tradition and of the medical and scientific communities.[64] Natural philosophers such as Marsilio Ficino, Pico della Mirandola and Pietro Pomponazzi supported an animated idea of nature and provided an authoritative theoretical justification for the widespread manipulation of 'the natural magic of objects'.[65] It is within the framework of this broader and expanding culture of prevention that the success of regimens for healthy living and the commitment of doctors to this genre should be understood.

The Distinctive Features of the Renaissance Regimen

The growing interest in preventive measures received fresh impetus from the state of emergency that characterized health conditions in late Renaissance Italy. During this period the ineffectiveness of medical treatment became fully evident in the face of recurrent epidemics, sustained warfare and frequent famines, and the arrival of the lethal and disfiguring French Disease. The principle that even the most serious illnesses were more likely to develop in a body that was imbalanced and unregulated reinforced the appeal of the doctrine of healthy living. According

62 Brian P. Copenhaver, 'Magic', in Daston and Park, *Cambridge History of Science*, pp. 518–40.

63 Jacqueline M. Musacchio, 'Conception and Birth', in Ajmar-Wollheim and Dennis, *At Home in Renaissance Italy*, pp. 124–35 and 'Lambs, Coral, Teeth, and the Intimate Intersection of Religion and Magic in Renaissance Italy', in Scott B. Montgomery and Sally J. Cornelison (eds), *Images, Relics, and Devotional Practices in Late Medieval and Renaissance Italy* (Tempe, AZ: Arizona Center for Medieval and Renaissance Studies, 2005), pp. 139–56.

64 On the enthusiasm of the medical community for amulets see Eamon, 'Markets', p. 219. On the links between learned and popular magical traditions in this period Federico Barbierato, *Nella stanza dei circoli: Clavicula Salomonis e libri di magia a Venezia nei secoli XVII e XVIII* (Milan: S. Bonnard, 2002).

65 Copenhaver, 'Magic', p. 520. On the idea of nature underpinning these practices see Lorraine Daston and Katharine Park, *Wonders and the Order of Nature 1150–1750* (New York, NY: Zone Books, 1998), Ch. 4.

to Carole Rawcliffe, a similarly troubled scenario, characterized by 'famine, environmental crisis, pestilence, demographic upheaval and social unrest', had stimulated a first wave of preventive literature between 1200 and 1500.[66] In the next two centuries the spread of print and the popularization of medical literature helped to disseminate this corpus of rules to a much larger audience; in the process, the preventive paradigm did not remain unchanged, as it is sometimes suggested, and many elements distinguish sixteenth and seventeenth-century regimens from their predecessors.[67] Though the full examination of these innovations would require much more space than is available here it seems worth outlining in these concluding pages some general trends that further reveal the vitality of the preventive discourse in the early modern period.

One striking difference that marks out Renaissance regimens from earlier examples of the genre is the amount of detail concerning the material and practical implementation of the advice. Recommendations become longer and more comprehensive, specific rather than generic, less abstract and general than in the medieval and ancient works. Like the books of secrets, books of healthy living acquire the form of practical guides. From the second half of the sixteenth century in particular, they make explicit reference to health-related domestic practices and everyday objects to be employed in health maintenance. For example, Durante does not simply mention, as do previous authors, the importance for preserving health of wearing appropriate clothes, he also details the type of fabric and fur these should be made of in winter and summer. Hence for the winter he recommends using clothes made of sheepskin or marten and socks made of hare skin (which help ailments of the joints), and fox skin (which comforts all the limbs); he further recommends covering hands with gloves in fox-skin, while in summer these should be made of kid skin or lambskin. When discussing healthy ways of sleeping he suggests placing on the stomach a small pillow filled with light, soft down, such as that of the vulture or, instead of a pillow, a little, fat dog. When discussing cloth he distinguishes between 'cloth for the head' and 'for the stomach', reflecting a tendency to differentiate items of household linen by usage that, as other sources confirm, had indeed already taken place in the late fifteenth century.[68] Similarly, he makes a distinction between ivory combs for the head and combs for the beard, warning that the two kinds should be kept separate.[69] As this evidence suggests the material culture of prevention was becoming more specialized and artefacts of daily use were charged with health significance.

Regimens also pay more attention to the domestic interior. In a period when the life of the elites became more centered on the home, the quality of the domestic

---

66 Rawcliffe, 'The Concept of Health', pp. 321–2.

67 For suggestions of immutability of the genre before the late seventeenth century, see Mikkeli, *Hygiene*, and Wear, 'History'.

68 Elizabeth Currie, 'Textiles and clothing', in Ajmar-Wollheim and Dennis, *At Home in Renaissance Italy*, pp. 342–51 (p. 344).

69 Durante, *Tesoro della sanità*, pp. 9–12.

environment clearly attracted increasing concern. Themes already present in ancient and medieval texts are given considerable space; regimens consider the location of the house, its exposure to different winds and to sun, and the materials it is built of; they also discuss the impact on health of architectural features such as the height and shape of ceilings, types of windows and their position, and the role of various objects, ranging from bed curtains to night caps, in protecting the body from harmful air.[70]

The redistribution of themes in the sixteenth-century regimen suggests that preventive advice was becoming more pervasive. It no longer focuses largely on diet, and the space devoted to food and its consumption diminishes even in comparison to fifteenth-century regimens, while the attention given to other non-naturals, and hence other spheres of life, expands considerably. Moreover, there is an expectation that readers will take care of their health more regularly, on a daily basis. Measures intended to purge the air of potentially dangerous vapours that were sometimes recommended by medieval regimens in time of pestilence or in particular months are now presented as quotidian operations to be performed as a matter of routine through the year.[71] As well as changing the air in the bedroom every morning, readers are therefore encouraged to burn fragrant woods in the fireplace in winter and to scatter flowers, aromatic herbs, vinegar and rose water in the rooms in summer.[72] Moreover, to limit the unhealthy effects of bad air, new techniques and devices are recommended that target the individual body more particularly. Regimens encouraged readers to inhale aromatic substances from little bags and sprinkle their garments and gloves with scented waters or powders. They also recommended specific scents for different parts of the body. So we learn that saffron, aloes and amber should be used to perfume protective cloths for the head and stomach; rosewater and vinegar were used in sponges worn on the chest and theriac or mithridate held in the mouth to correct the malignity of the air.[73] The frequent instructions, in books of secrets, for the preparation of

---

[70] See, for example, Bartolomeo Paschetti, *Del conserva la sanità* (Genoa, 1602), p. 132; Petroni, *Del Viver*, p. 295; Durante, *Tesoro della sanità*, p. 7; Tomaso Rangoni, *Alla serenissima signora Loredana Mocenica. come la Duchessa e i Venetiani possano vivere sempre sani* (Venice, 1570), p. 10.

[71] For medieval authors who associate the need to purify the air only with pestilence see Aldrobrandino da Siena (d. c. 1287), *Le Régime du Corps*, ed. L. Landouzy and R. Pepin (Paris: H. Champion, 1911), pp. 59–61 and *Opera Utilissima* (translation of a regimen by Maino de Maineri, d.1368), p. 104; others, like Taddeo Alderotti (d. 1295) do not include any air-related recommendations: *Libello per conservar la sanità del corpo fatto per maestro Taddeo da Firenze*, ed. Ignazio Galeati and F. Galeati (Imola, 1852).

[72] Pictorius, *Dialogi*, p. 23; Durante, *Tesoro della sanità*, pp. 4 and 6; Domenico Panaroli, *Aerologia cioè discorso dell'Aria, trattato utile per la sanità* (Roma, 1642), pp. 86–7 and Rangoni, *Alla serenissima signora Loredana Mocenica*, p. 10.

[73] Ficino, *Tre vite*, p. 13; Durante, *Tesoro della sanità*, pp. 6 and 10–11 and Bartolomeo Traffichetti, *L'arte di conservar la salute tutta intiera trattata in sei libri* (Pesaro, 1565), p. 226.

scented oils, waters and pastes to sprinkle or rub on the body and the clothes and to be burned in the fire or perfume-burner, and for making pomanders and little bags of herbs and flowers to be carried around confirm the popularity of these practices.

Early modern regimens also provided more precise guidelines about standards of hygiene. According to humoural theory, bodily hygiene was classified as a form of purging or 'evacuation' of superfluities that built up in various parts of the body during the processes whereby food was turned into physical matter. If this residual waste was not removed it could severely damage the body, causing blockages, fevers, debilitations and other evils. All types of body emissions were therefore regarded positively in humoural physiology, so long as they were not protracted and did not endanger the natural heat of the body. In the Renaissance, however, the emphasis shifts away from the preventive use of the most violent evacuations, namely vomit, enema, sweating and blood-letting, while there is considerably greater attention devoted to the performance of daily hygiene. The list of organs to be regularly purged increases, and the operations that constitute the daily morning toilette become far more detailed.[74] This growing emphasis on hygienic practices centered on the head (seat of the brain, and hence of the rational faculty) tallies with the recurrent presence of recipes for the manufacture of various types of waters, toothpastes and toothpicks found in the books of secrets. Even combing the hair and the beard is presented in regimens as a healthy practice rather than a ritual of beautification.

Although some of these recommendations were already present in medieval regimens, they were much less detailed and were designed to preserve the functions of the organs in questions – speech, hearing and sight.[75] In the Renaissance texts became more focused on removing dangerous substances, now obsessively described as degenerating 'excrements'. Hair and beards were therefore to be kept short so as to enable the constant expulsion of waste, especially from the scalp.[76] Moreover, while the medieval regimen did not usually specify the frequency with which hygienic operations were to be performed, Renaissance regimens prescribe how often the face, hands, feet and head should be washed and even how often shirts, socks and bed-linen should be changed.[77] Another new feature of the culture of prevention of this period was the importance attributed to personal and

---

74 For an explicit criticism of the use of blood-letting and vomit in any part of the year: Lemnius, *Complessione del corpo*, p. 40.

75 For example, in *Opera utilissima*, a translation of a fourteenth-century regimen, the cleaning of the tongue speeds speech and that of the ears removes superfluities that impair the hearing, pp. 23–5 and 29 and Alderotti, *Libello*, pp. 6, 7 and 12.

76 Lemnius, *Complessione del corpo*, p. 98; Luc'Antonio Camaffi, *Reggimento per viver sano nei tempi caldi* (Perugia, 1610), p. 240 and Viviani, *Trattato del custodire la sanità*, pp. 124–5.

77 Rangoni, *Alla serenissima signora Loredana Mocenica*, p. 10 and Fonseca, *Del conservar la sanità*, pp. 16 and 101.

household linen as a hygienic tool for absorbing the sweat and grime expelled from within the body.

Through these detailed guidelines, doctors endeavoured to regulate an increasing number of the activities normally performed in domestic life. By providing punctilious instructions about the operations of eating, sleeping, exercising and cleansing of the body, they not only stressed their own authority and knowledge, but equipped lay people to look after their bodies. In a period rife with health hazards, doctors offered suggestions that enabled people to play an active part in maintaining their well-being rather than remaining passive and without protection. For example, by encouraging the practice of extensive daily evacuations through various forms of hygienic care, in place of violent evacuations performed in response to ailments and plethora, they provided patients with a degree of control over the degenerative processes that could lead to disease. Moreover, by stressing the health significance of the domestic environment and the objects populating the Renaissance house, the authors of these texts considerably expanded the domain of what could be considered preventive and hence affected by human action.

Conclusions

The enabling and reassuring function performed by the adoption of daily preservative and preventative measures helps to explain the success of regimens of health in the age of print, and of their companion genre, the books of secrets. The latter also included many remedies that were preventive, rather than therapeutic, especially recipes for assisting in the hygienic operations required by a body, which was increasingly represented as continuously excremental and hence in need of regular cleansing. While the focus on the popularity of therapeutic recipes and their intense circulation has led to the conclusion that early modern patients had developed a new approach to illness, rejecting regimens of health and simply looking for remedies to relieve their symptoms, the present investigation offers a more balanced picture, suggesting that great care was taken to keep the body in good shape as well as to heal it when it fell ill. The two approaches are not to be seen as alternative and mutually exclusive. As well as being consumers of all sorts of medicaments, Renaissance people were also voracious consumers of preventive advice; the care of the healthy body coexisted with the recourse to medicinal remedies when things went wrong in the same way as domestic medicine survived the spread of commercial medicine and for a long time ran in parallel with it.[78] The printing strategies of authors, editors and translators of regimens, the fact that they continued to release new titles and modified editions of the same texts, their use of vernacular and small format editions, and the extension of the contents of these handbooks to a range of matters, well beyond the regulation of food, all demonstrate the existence of a lively market for preventive advice. On the other

[78] Leong, 'Making medicines' and Leong and Pennell, 'Recipe Collection'.

hand, the sensitivity of this literature to changing lifestyles and fashions and to new features of the material world provides an additional key to understanding the appeal that early modern regimens exerted on a public that would find a familiar picture in their graphic descriptions.

Chapter 10
Secrets of Place: The Medical Casebooks of Vivant-Augustin Ganiare

Lisa Wynne Smith

Vivant-Augustin Ganiare, physician of Beaune in Burgundy, kept a detailed account of his busy medical practice between 1736 and 1777. As a frame for his medical observations, he maintained a record of the monthly weather and diseases, as well as the state of the local crops. In particular, he had much to say about his patients' failings. Typical of his commentary is a complaint about intemperance:

> We have said more than once that of all the patients entrusted to us by Providence, there were hardly any who gave us more grief in Practice, and who were less a credit to the doctor, than the persons subject to wine.[1]

Ganiare's criticism summarizes three of his regular themes: the religious context of his practice, the inability of patients to listen and the problems of alcohol. Historians familiar with Ganiare's casebooks have been struck by his judgmental tendencies; 'always one to blame the patient not the physician for failure',[2] Ganiare 'vigorously denounced certain aspects of his contemporaries' way of life'.[3] But they have overlooked the meaning of those tendencies, which are what make the casebooks so compelling: what was the significance of his frequent disapproval?

There are three parts to the answer, which cannot be easily disentangled. First, the voice of condemnation was central to his textual authority – his moral advantage allowed him to be a reader of secrets (whether of God, Nature or human nature), which he chose to share for the good of young physicians, as well as

[1] 'Nous avons dit plus d'une fois que de tous les malades que la Providence nous confioit il n'y en avoit point qui nous donna plus de peine dans la Pratique, et qui fit moins d'honeur au medecin, que les personnes sujettes au vin'. Bibliothèque Municipale, Dijon (hereafter BM, Dijon), p. 459.

[2] Laurence Brockliss and Colin Jones, *The Medical World of Early Modern France* (Oxford: Oxford University Press, 1997), p. 571.

[3] 'Dénonce vigoureusement certains aspects du mode de vie de ses contemporains'. Christine Lamarre, 'Observations et Reflexions sur le temps qu'il fait au dix-huitième siècle: les manuscrits du Docteur Ganiare, de Beaune (Borgogne), 1736–1777', in Lucette Davy (ed.), *Météorologie et catastrophes naturelles dans la France méridonale à l'époque moderne* (Montpellier: Université Paul Valery III, 1993), p. 236.

his patients. Second, his moral advantage was derived from his religious belief and scientific methodology, which in turn shaped his understanding of patients and medicine. Finally, the Enlightenment provided a broad intellectual context, emphasizing the need for self- and social management, especially in relation to men. Ganiare's medical practice provides an excellent case study of how these ideas played out in daily life in a provincial French town, and his casebooks can be read as books of secrets. Through his casebooks, he revealed to interested patients and physicians a simple formula for healthy living, as well as uncovering the inner workings of Beaune's environment and society. Ganiare derived his secrets for healthy living from his observations of weather and disease, closely connecting them to his patients' place – their local environment, Burgundian locale and social hierarchy – and filtering them through his own his religio-scientific worldview and privileged background.

Ganiare and His Casebooks

Ganiare (1698–1781) was born in Beaune, a Burgundian town of 10,000, to an elite family; his father was Secrétaire du Roi of Dole and his mother came from a local medical family. Ganiare took his degree in Montpellier and started practicing in 1736.[4] He worked at the Beaune Hôtel-Dieu and had a large private clientele in which he saw between 14 and 50 patients daily.[5] In 1765, the height of his career, he treated 2,216 patients. Even in his last year of practice, aged 79, he had 800 patients. For 40 years, he maintained detailed notes of his practice, comprising 12 books (5,500 pages).[6] The first casebook (1736–43) was in Latin, but when his practice rapidly expanded following the death of another physician, Ganiare found that he only had time to write in French.[7] Each month, he outlined the weather trends, unusual behaviour amongst the populace, and reigning diseases before discussing specific cases, organized according to disease type. Mostly, he treated a narrow range of problems, from colds and fevers to rashes and smallpox, using a combination of old (Hippocrates) and new (Thomas Sydenham) methods.[8]

In many ways, Ganiare is a good example of a middling physician. He had a thriving provincial practice, even if he never attained the heights of fame and fortune, and had an important library in Beaune. He was indeed a well-read physician, familiar with the works of non-French physicians.[9] Nearby Beaune,

[4] Georges Chevaillier, *Médecine et médecins à Beaune: des origines au XXè siècle* (Beaune: Centre Beaunoise d'Études Historiques, 2004), p. 54.

[5] Brockliss and Jones, *Medical World*, p. 535n.

[6] Lamarre, 'Observations', pp. 231–2.

[7] BM Dijon, MS 426, p. 1.

[8] Brockliss and Jones, *Medical World*, pp. 571, 573n.

[9] Chevaillier, *Médecine*, p. 54; Brockliss and Jones, *Medical World*, p. 573n; BM, Dijon, MS 425, Ephemeredes of Vivant-Augustin Ganiare, pp. 5–10.

the active Dijon Academy (established 1741) was particularly interested in the importance of natural philosophy for medical practice.[10] Historians Brockliss and Jones have described the learned French physician of the mid-eighteenth century, who was to be fluent in Latin and French, as well as able to read other European languages in order to keep up with medical developments: he should keep good notes of his practice and study the weather to develop medical science. In addition, he would provide medical charity and be an upstanding member of the community.[11] Ganiare, provider of medical charity and good citizen of Beaune, was such a physician. In this chapter, I focus only on the casebooks for 1745 through 1749.[12] By this time, Ganiare had an established practice in Beaune and he had settled into his roles of physician and man of science. His casebooks from this time provide insight into the worldview of a well-established mid-career provincial physician, as well as the patients and community of Beaune.

It is unusual to read eighteenth-century medical casebooks as books of secrets, but Ganiare was keenly interested in observing God and Nature, with the intention of uncovering their secrets for the betterment of society. Indeed, he even framed his casebooks in terms of secrets. In an entry from January 1749, Ganiare promised to divulge to his readers his failures in medical practice, which ordinarily would remain unseen:

> My intention is not to critique my confreres, still less to blame the Practice that they put into usage; I have neither superior talents to set myself up as dictator, and to make shine my knowledge which is only pure ignorance … [B]ut solely to teach young practitioners to avoid falling into the same faults [as mine] if by chance they should find themselves in similar situations.[13]

At his death, Ganiare bequeathed his casebooks to the town of Beaune rather than to an individual.[14] He, perhaps, would have seen it as typical of the townspeople's rejection of his advice that the town sold the books to Dr Ripault, a Dijon physician, 30 years later.

---

[10] Daniel Roche, 'Natural History in the Academies', in N. Jardine, J.A. Secord and E.C. Spary (eds), *Cultures of Natural History* (Cambridge: Cambridge University Press, 1996), pp. 133–5, 141.

[11] Jansenists also encouraged charity. Brockliss and Jones, *Medical World*, pp. 285, 474; Dale Van Kley, *The Jansenists and the Expulsion of the Jesuits from France 1757–1765* (New Haven, CT and London: Yale University Press, 1975), p. 10; Lamarre, 'Observations', p. 232.

[12] BM, Dijon, MSS 426–7.

[13] 'Mon dessein n'est pas de critiquer mes confreres, encore moins de blâmer la Pratique qu'ils mettent en usage; je n'ai ni des talens superieurs pour m'eriger en dictateur, et faire briller mon sçavoir qui n'est que pure ignorance … [M]ais uniquement pour apprendre aux jeunes practiciens a ne pas tomber dans les memes fautes si par hazard ils se trouvent dans des cas semblables'. BM, Dijon, MS 426, p. 123.

[14] Lamarre, 'Observations', pp. 231–2.

Historians of secrets have tended to overlook the eighteenth century, but Ganiare's casebooks suggest that the tradition may have continued in different forms shaped within an eighteenth-century context. A profoundly religious Jansenist (a sect of Catholicism that emphasized man's corruption and lack of free will), Ganiare believed that God's will could be seen operating in Nature. He was also a good Enlightenment physician, interested in natural history and its methodical study. For Ganiare, it was possible to find the secrets for healthy living, the cure and prevention of disease, by reading the signs that God had placed in Nature, such as weather and illness progression. Each month, he recorded the local weather and local habits, indicating what diseases predominated and detailing cases of individual patients. His goal: to understand the secrets of Nature and their effects on the townspeople. Ganiare's social status as a member of an elite Beaunois family was also important which, along with his pessimistic religious views, predisposed him to assume that men who lacked self-control were sources of social disorder. In many ways, Ganiare's concerns were typical of contemporary French medical thought. During the eighteenth century, the ability of a man to control his body and emotions, as well as to govern others, was of growing political importance. Ganiare scrutinized his patients' behaviour, particularly when a man failed to fulfil the responsibilities attached to his place within the social order. Not only did he read men's characters through their diseases, but he believed that local weather and epidemics exposed the disruption of Beaune's social order.

Books of Secrets?

As many historians have discussed, secrets were coupled with ideas about Nature and scientific knowledge in pre-modern society. Secrets were mysterious, to be concealed even by those who had learned them.[15] This applied to diverse areas, such as natural philosophy, anatomy, alchemy, magic, women's bodies and charlatans' remedies.[16] Published books of secrets as a genre started to fall out of favour during the seventeenth century, supplanted by books of knowledge. The earlier focus on secrets highlighted a possessor's special ability to manipulate the natural

---

15 William Eamon, *Science and the Secrets of Nature: Books of Secrets in Medieval and Early Modern Culture* (Princeton, NJ: Princeton University Press, 1994), pp. 48, 58 and 81.

16 Eamon, *Science*; Susan Broomhall, '"Women's Little Secrets": Defining the Boundaries of Reproductive Knowledge in Sixteenth Century France', *Social History of Medicine* 15/1 (1999): pp. 1–15; Monica Green, 'From "Diseases of Women" to "Secrets of Women": The Gynecological Literature in the Later Middle Ages', *Journal of Medieval and Early Modern Studies* 30/1 (2000): pp. 5–40; Katharine Park, *Secrets of Women: Gender, Generation, and the Origins of Human Dissection* (New York, NY: Zone Books, 2006). On charlatanism and secrecy, see David Gentilcore, *Medical Charlatanism in Early Modern Italy* (Oxford: Oxford University Press, 2006).

world, but a later concern with knowledge emphasized that the natural world was predictable and manageable once the natural laws were understood.[17] Eighteenth-century natural philosophers wanted to share their knowledge of Nature's secrets with others, with the aim of achieving a comprehensive rather than specialized knowledge of the world. On an individual basis, the quest for knowledge required disciplined and sustained study of natural phenomena: experimenting, observing and recording.[18] Only certain types of people were thought to have the capacity to learn and to use secrets, defined as much by one's class and gender, as personal qualities of reliability, intelligence, curiosity, moral and physical self-control.[19] Some historians have considered the ways in which the discourse of uncovering secrets and scientific knowledge was strongly masculine and increasingly violent, moving from depictions of the gentle unveiling of Nature in the Middle Ages to the penetration of Nature by the seventeenth century.[20] Significantly, as all of these studies have shown, issues of control, whether of Nature, of knowledge or of one's self, are at the heart of any discussion of secrets.

There are, however, religious, temporal and gender gaps in the existing historiography. Religion, for example, is central to the discussion of medieval secrets,[21] but early modernists have considered religion and secrets in little detail.[22] Indeed, there have been relatively few studies on the relationship between early modern medicine and religion, despite the recognition that it was important.[23]

---

17 Alison Kavey, *Books of Secrets: Natural Philosophy in England, 1550–1600* (Urbana, IL and Chicago, IL: University of Illinois Press, 2007), pp. 157–8.

18 Lorraine Datson and Peter Galison, *Objectivity* (New York, NY: Zone Books, 2007); Eamon, *Science*, pp. 292–99; Kavey, *Secrets*; Adi Ophir and Steven Shapin, 'The Place of Knowledge: A Methodological Survey', *Science in Context* 4/1 (1991): pp. 3–21, 9; Steven Shapin, *A Social History of Truth: Civility and Science in Seventeenth-Century England* (Chicago, IL: University of Chicago Press, 1994).

19 Kavey, *Secrets*, pp. 62–4, 74–7, 83; Shapin, *Truth*, pp. 65–125.

20 Ludmilla Jordanova, *Nature Displayed: Gender, Science and Medicine, 1760–1820* (London and New York, NY: Longman, 1999); Evelyn Fox Keller, *Secrets of Life, Secrets of Death Essays on Language, Gender and Science* (London and New York, NY: Routledge, 1992), pp. 56–72; Carolyn Merchant, *The Death of Nature: Women, Ecology, and the Scientific Revolution* (San Francisco, CA: Harper and Row, 1980); Londa Schiebinger, *Nature's Body: Gender in the Making of Modern Science* (Boston, MA: Beacon, 1993).

21 Eamon, *Science*; Park, *Secrets of Women*.

22 Those who discuss it have largely focused on English Protestantism: Eamon, *Science*; Keller, *Secrets*, pp. 56–72.

23 Cf. Ole Peter Grell and Andrew Cunningham (eds), '*Religio Medici': Medicine and Religion in Seventeenth-Century England* (Aldershot: Ashgate, 1996); Andrew Wear, *Knowledge and Practice in English Medicine, 1550–1680* (Cambridge: Cambridge University Press, 2000); Laurence Brockliss, 'The Medico-Religious Universe of an Early Eighteenth-Century Parisian Doctor: The Case of Philippe Hecquet', in Roger French and Andrew Wear (eds), *The Medical Revolution of the Seventeenth Century* (Cambridge: Cambridge University Press, 1989), pp. 191–221.

Religion, instead, often appears at the margins of the history of medicine, discussed as a convention or catch-all explanation.[24] Ganiare's interpretation of natural phenomena and illness was founded upon his Jansenist religious beliefs.[25] Jansenism was a sect of Catholicism that emphasized man's complete corruption and lack of free will. Ganiare's insistence that patients take responsibility for their own health was bolstered by his religious beliefs, which demanded strict behaviour and morals to avoid worldly distractions. Despite its negativity about human nature, Jansenism was deeply optimistic and practical, stressing that devotion and charity could make a better society.[26] An analysis of the casebooks provides a French case study of how one physician's religious worldview shaped his medical practice.

The temporal context is also important, with little work undertaken on the fate of secrets in the eighteenth century.[27] Traces can be found in eighteenth-century medicine and natural philosophy, with people wielding increasing control over the natural world. Much was at stake; once Nature's mechanisms were known, those who had uncovered the secrets could control Nature. This had implications for medical thought and practice, especially given the eighteenth-century goals of building a healthy nation and managing the body politic.[28] European physicians believed that they could discover the causal relationship between environment and epidemics and recorded meteorological details alongside medical notes to identify correlations. In 1767, Jean Razoux (*Tables nosologiques & météorologiques*)

---

24 Cf. Brockliss and Jones, *Medical World*; Mary J. Dobson, *Contours of Death and Disease in Early Modern England* (Cambridge and New York, NY: Cambridge University Press, 1997), pp. 268–270; Shapin, *Truth*, pp. 63–4, 181–2 and 209–10.

25 Lamarre found textual evidence indicating that Ganiare may have been a Jansenist: gratitude on the expulsion of the Jesuits, the enemies of the Jansenists, from France. A comparison of Ganiare's religious and medical ideas (based on the casebooks) with those of Philippe Hecquet, a known Jansenist physician in the early eighteenth century, shows a substantial overlap. Lamarre, 'Observations', p. 232; Brockliss, 'Medico-Religious', pp. 191–221.

26 Brockliss, 'Hecquet', pp. 194–5, 201, 206–8 and 213; Van Kley, *Jansenists*, pp. 10 and 34; Alexander Sedgwick, *Jansenism in Seventeenth-Century France* (Charlottesville, VA: University Press of Virginia, 1977), pp. x–xii; Brockliss and Jones, *Medical World*, p. 370.

27 The focus for eighteenth-century secrets has been on charlatans. Cf. Gentilcore. *Charlatanism*.

28 Marie-Noëlle Bourguet, 'Measurable Difference: Botany, Climate, and the Gardener's Thermometer in Eighteenth Century France', in Londa Schiebinger and Claudia Swan (eds), *Colonial Botany: Science, Commerce, and Politics in the Early Modern World* (Philadelphia, PA: University of Pennsylvania Press, 2005), pp. 270–86; Merchant, *Death*, pp. 188–90; Schiebinger, *Nature's Body*; Emma Spary, '"Peaches, which the Patriarchs Lacked": Natural History, Natural Resources, and the Natural Economy in France', *History of Political Economy*, Annual Supplement 38 (2003): pp. 14–41; Emma Spary, 'Political, Natural and Bodily Economies', in Jardine and Secord (eds), *Cultures*, pp. 178–96.

outlined the necessary information: a locale's terrain, climate, weather, character of citizens, diet and regular illnesses. Ideally, practical medical improvements would result from cumulative experience and observation.[29] Certainly, in an age of state-building and colonization, the ability to exploit the New World by managing 'new' diseases and environments or understanding local remedies was of paramount importance.[30] Ganiare's casebooks brought together old and new scientific traditions, framing his secrets according to the new scientific terminology: knowledge that he had uncovered, but would share for the common good.

The eighteenth century was also a key period for the history of gender in terms of establishing gender roles and political claims.[31] During the eighteenth century, manly self-control was important in both politics and science. Men of the middling sorts and urban elite sought gains in (or consolidation of) political power. Manly independence, central to their claims of political virtue, was demonstrated by their public and private honour.[32] This in turn was inscribed on their bodies, with ideal masculine bodies reflecting bourgeois principles of self-control and containment. A male body that did not behave appropriately, particularly if weakened through a man's own behaviour, suggested a lack of discipline.[33] Independence was integral to the practice of science, too. Since some things could not be studied directly

29 James Riley, *The Eighteenth-Century Campaign to Avoid Disease* (New York, NY: St Martin's Press, 1987), pp. x, xi, 9–10, 33 and 45; Gordon Manley, 'The Weather and Diseases: Some Eighteenth-Century Contributions to Observational Meteorology', *Notes and Records of the Royal Society of London* 9/2 (1952): pp. 300–307.

30 James McClellan and François Regourd, 'The Colonial Machine: French Science and Colonization in the Ancien Regime', *Osiris*, 2nd series, 15 (2000): pp. 31–50; Mark Harrison, 'From Medical Astrology to Medical Astronomy: Sol-lunar and Planetary Theories of Disease in British Medicine, c. 1700–1850', *The British Journal for the History of Science* 33/1 (2000): pp. 25–48.

31 Karen Harvey, 'The History of Masculinity, ca. 1650–1800', *Journal of British Studies* 44/2 (2005): pp. 296–312; Thomas Laqueur, 'Sex in the Flesh', *Isis* 94/2 (2003): pp. 300–306; Londa Schiebinger, *The Mind Has No Sex?: Women in the Origins of Modern Science* (Cambridge, MA: Harvard University Press, 1989).

32 Anna Clark, 'The Chevalier d'Eon and Wilkes: Masculinity and Politics in the Eighteenth Century', *Eighteenth Century Studies* 32/1 (1998): pp. 19–48; Sean Quinlan, 'Men without Women? Ideal Masculinity and Male Sociability in the French Revolution, 1789–99', in Christopher E. Forth and Bertrand Taithe (eds), *French Masculinities: History, Culture and Politics*, (Basingstoke: Palgrave, 2007), pp. 31–50; Anne C. Vila, 'Elite Masculinities in Eighteenth-Century France', in Forth and Taithe (eds), *French Masculinities*, pp. 15–30.

33 Norbert Elias, *The Civilizing Process*, (1994, repr. Oxford, Wiley-Blackwell, [2000]), pp. 109–21; Elizabeth Green Musselman, *Nervous Conditions: Science and the Body Politic in Early Industrial Britain* (Albany, NY: State University of New York Press, 2006), p. 6; Michael Stolberg, 'An Unmanly Vice: Self-Pollution, Anxiety and the Body in the 18th Century', *Social History of Medicine* 13/1 (2000): pp. 1–21; Fernando Vidal, 'Onanism, Enlightenment Medicine, and the Immanent Justice of Nature', in Lorraine

by all scientists (like odd weather), witness accounts were crucial and needed to be trustworthy. But only certain types of people – those of higher social status, learned occupation and male gender – could claim authority and reliability.[34] Scientific methodology revealed the underlying patterns or secrets of Nature, but its demands of literacy and reliability ensured that those ignorant of them could never claim mastery over Nature. Ganiare, a physician and Jansenist from the urban elite, judged his male patients according to these high standards. In contrast to their deficient behaviour, he was the ideal man to uncover and to share secrets. By examining Ganiare's casebooks within their religious, temporal and gender contexts, I provide a case study of how one eighteenth-century provincial French physician framed his medical practice in terms of 'secrets'.

The Secrets of God, Nature and Medicine

The relationship between God, Nature and medical secrets provided the overarching narrative framework in Ganiare's casebooks. For Jansenists, mankind's relationship with Nature was fraught with danger. After the Biblical Fall, man had become susceptible to disease and death, while the environment became tainted and unhealthy.[35] Nature was hostile, and God was able to wield Nature either to punish or to show his mercy. But being loving, as well as stern, God had placed remedies to all of man's ills in Nature. By reading Nature's Book of Secrets, mankind could uncover the secrets to living a long and healthy life. The best reader of Nature was ideally a physician, someone who understood the progress of disease and the operations of the human body.[36] Jansenists, moreover, believed that God was hidden, only occasionally choosing to reveal himself through signs. When any evidence of God did appear, it could be taken as proof of his everlasting existence.[37]

Datson and Fernando Vidal (eds), *The Moral Authority of Nature*, (Chicago, IL and London: University of Chicago Press, 2004), pp. 254–81.

34 Vladimir Jankovic, *Reading the Skies: A Cultural History of English Weather 1650–1820* (Chicago, IL: University of Chicago, 2000), p. 98; Robert A. Nye, 'Medicine and Science as Masculine "Fields of Honor"', *Osiris*, 2nd series 12 (1997): pp. 60–79; Shapin, *Truth*; Shapin and Simon Schaffer, *Leviathan and the Air Pump: Hobbes, Boyle and the Experimental Life* (Princeton, NJ: Princeton University Press, 1985), p. 62.

35 Andrew Wear, 'Making Sense of Health and the Environment in Early Modern England', in Andrew Wear (ed.), *Medicine in Society: Historical Essays* (Cambridge: Cambridge University Press, 1992), pp. 140–41.

36 Brockliss, 'Medico-Religious', pp. 206–8.

37 Kreiser discusses the importance of miracles and Providence to eighteenth-century Jansenists. Hecquet was initially willing to believe in the miraculous claims of the *convulsionnaires* in Paris, although he later argued that the convulsions were natural rather than miraculous. Brockliss, 'Medico-Religious', pp. 216–18; B. Robert Kreiser, *Miracles, Convulsions, and Ecclesiastical Politics in Early Eighteenth-Century Paris*

These ideas clearly emerge throughout Ganiare's casebooks. In the entry at the start of MS 427 (January 1749), for example, Ganiare reflected on how fevers were common, but mysterious. God, he argued, had not made such secrets impenetrable, even if we could not yet understand them: by reading fevers, 'it is by observation that we are instructed in the variety of paths that it clears to conduct an illness to its cure'.[38] Knowledge, moreover, was not immediate, needing to be accumulated by the physician over time: 'he needs a long experience upheld by a multitude of facts, of which a union forms a whole'.[39] Thus, a physician should always experiment with different approaches, since even failures could add to knowledge. More importantly, physicians should diligently record everything. 'Science', he wrote:

> consists of observing it [Nature] closely to examine with care its variations, movements, progress, the different faces by which it presents the routes that she follows to make and to perfect the great work of purifying[40]

According to Ganiare, perseverance and the methods provided by science would ultimately decipher God's messages. This had immediate applications for medicine, as well: by learning to differentiate fevers, the most common of problems, physicians would best help their patients. Secrets, to Ganiare, were God's messages left for man's benefit, to be deciphered by the skilled practitioner of science.

Just as Ganiare saw fever as one of God's signs, he identified God's hand in the operation of local weather even when famine threatened. For eighteenth-century England, historian Vladimir Jankovic has argued that meteorologists believed that weather made visible the divine and that their work was to interpret signs.[41] In 1749, a difficult year with regular threats of famine, Ganiare read in the strange weather God's mercies to the poor, believing that God had prevented famine solely out of care for the poor. The trouble started in February 'with rain, which became so abundant and continued so long that the rivers burst their banks'. Floods in low-lying areas made people fear damage to the villages – 'but

---

(Princeton, NJ: Princeton University Press, 1978), pp. 70–73; Steven Nadler, *Arnauld and the Cartesian Philosophy of Ideas* (Manchester: Manchester University Press, 1989), pp. 136–8; Sedgewick, *Jansenism*, pp. 48–9.

[38] 'C'est par observation que nous sommes instruits de la varieté des Routes qu'elle [Nature] se fraye pour conduire une maladie a sa guerison'. BM, Dijon, MS 427, p. 915.

[39] 'Il a besoin d'une longue experience soutenüe d'une multitude de faits dont la reunion formee un tout'. Ibid., p. 916.

[40] 'Consiste à l'observer de prez d'en examiner avec soin les variations, les mouvemens, les progrez, les faces differentes sous lesquelles elle [Nature] se presente les routes qu'elle suit pour faire et perfectioner le grand ouvrage de la depuration'. Ibid., p. 916.

[41] Jankovic, *Reading*, pp. 129–38.

Divine Providence did not permit (in favour of the poor) a tragedy that would necessarily be followed by a Great Famine'.[42] A bitter cold snap at the end of the month brought the famous winter of 1709 to everyone's mind, but Providence ensured that no damage was done that month.[43] In mid-May, the vine crop was almost entirely lost when the weather turned wintry and locals were so worried that that they had a series of public prayers. 'Happily', however, 'Providence always attentive to the needs of the Poor did not allow their hope to be misled' and preserved their crops.[44]

June was one of the hardest months in years: 'the most unfortunate, most sad, and most dangerous'.[45] The month started off excessively hot and dry, but then became so cold that people lit their fires. More vines were lost and, by the end of the month, a shortage of wheat had 'the poor suffering, muttering, complaining loudly, the rich started to fear for themselves'.[46] Money stopped circulating and they awaited once more the Great Famine, especially when the vines froze for the third time on June 29. Again, crisis was avoided when July proved hot enough to ripen the crops.[47] But the heat was a mixed blessing. It was so hot that by August that vapours were rising from the earth, causing lung ailments. Once more, divine intervention ensured the safety of Beaune: 'Divine Providence ceased (for a time) this frightening malady' by making the air cooler and moister.[48] For the rest of 1749, the weather remained (happily for Beaune) uneventful. Despite the unusually harsh weather from February to August 1749, the worst consequences of the floods, cold and heat had been repeatedly averted at the last moment. Providence, according to Ganiare's monthly reports, intervened each time to ensure that the poor did not suffer from what easily could have been a terrible famine. To Ganiare, the signs in the weather were clear: God was merciful. Through the study of patterns in Nature, it was possible to uncover the messages that God had left for the careful and faithful, whether it was a reminder of his manifold mercies or secrets for conquering disease.

---

42 'Avec la pluye, qui devint assez abondante, et continua assez long-tems pour faire deborder les rivieres … mais la Divine Providence ne permit pas (en faveur des pauvres) un malheur qui eut été suivi necessairement d'une Grand Famine'. BM, Dijon, MS 427, p. 935.

43 The extreme cold that had descended upon Europe between December and March of 1709 interested other physicians, such as Hoffman and Ramazzini. Riley, *Campaign*, p. 33.

44 Heureusement la Providence toujours attentive aux besoins des Pauvres ne permit pas que leur esperance fut trompée'. Ibid., pp. 1004–5.

45 'Le plus facheux, le plus triste, le plus dangereuse'. Ibid., p. 1026.

46 'Les pauvres souffroient, murmuroient, crioient, les riches commençoient a craindre pour eux memes'. Ibid., p. 1026.

47 Ibid., pp. 1046–7.

48 'La divine Providence fit cesser (pour un temps) cette effrayante maladie'. Ibid., p. 1067.

Secrets of Healthy Living for the Beaune

Secrets were not only concealed within the weather or environment, but could be shown in the relationship between place and disease. Ganiare considered the contributions of weather, regional industry and local character to his patients' illnesses. The health of a population was equated with the health of their environment, with each environment, like individual people, having a particular temperament.[49] By the eighteenth century, physicians like Arbuthnot, Ramazzini and Hoffman studied the prevailing diseases of an area to specify what diseases 'reigned' locally and at what time of year.[50] This connects to what Jankovic sees as an eighteenth-century meteorological discourse about 'nature of place and the place of nature' that connected geography to issues of regional and national identification.[51] Certainly, Ganiare's lifelong study of health and the environment considered only the area in and around Beaune, known for its erratic weather. By considering his patients' illnesses within their geographic and climactic context, Ganiare was able to offer suggestions for how best to prevent those diseases – but patients were not always willing to take his advice. Ganiare's accounts of the local area presented two types of secrets: how to live healthfully in the Beaune area and what patients' ill health revealed of their characters.

Given that Beaune was in the heart of wine country (Burgundy), it is not surprising that the dangers of the vine play an important role in Ganiare's world view. When wine was not directly causing ill health, the environment in which patients lived became the problem. During winter, the winds were cold, which meant that men drank excessively to stay warm, people failed to dress appropriately, and still others kept their houses too hot. Summer was no better, since people slept with their windows open, men caught chills from moving between the heat outside and the coldness of wine caves, and women did not wear enough clothes. Ganiare regularly noted that the local weather was subject to sudden shifts in temperature at any time of the year. For example, Ganiare complained that December 1748 was so warm and humid that the youth did not cover their heads 'under the pretext that it was not cold enough'.[52] But then the temperature suddenly turned glacial and those who had not worn their hats suffered from colds and chest ailments. Ganiare constantly reiterated the need for careful dressing, as he did in March 1749. The days were so warm as 'to put the imprudent to the occasion of leaving off some of their clothes: the women above all reduced the number of their skirts'. People, he complained, should have worried more about 'a season whose dominant character is inconsistency' because the sudden return to wintry temperatures near

---

49 Dobson, *Contours*, pp. 51–60 and 287–336; Wear, 'Making Sense', pp. 126–36.

50 Riley, *Campaign*, pp. 11–12 and 33.

51 Jankovic, *Reading*, p. 5.

52 'Sous pretexte qu'il ne faisoit pas assez froid'. BM, Dijon, MS 427, p. 917.

the equinox led to fevers and colds.[53] Ganiare's first secret for healthy living was simple: dress appropriately for the changeable weather.

The regional wine-making industry provided the background for many illness narratives: patients contracted illnesses when wine-making, working in wine cellars or harvesting grapes. Ganiare's clientele largely consisted of middling and elite local families, particularly those connected to the wine trade. Significantly, although the consumption of alcohol played little role in Ganiare's discussion of women's health, it was a key interest for him when discussing his male patients. According to Ganiare, drinking too much regularly was most dangerous, wearing one's constitution down over time to ruin it entirely. To give but one example of the problems that wine could cause, consider the case of 'a man of 47 years, very stout, accustomed to eating well', whose business matters often obliged him to drink more than agreed with his temperament. One time, the patient overheated after drinking too much and then caught cold while walking home in the glacial north wind. At home, he drank mulled wine to warm up, but forgot to change his linens before falling asleep. The result: he awoke with no voice.[54] Others did not even use a business excuse for drinking too much, such as a bourgeois over the age of 60 who was predisposed given to drinking wine and 'hostile enough to exercise'.[55] Tormented for three weeks by a cough, he became less robust than usual, with legs so weak that walking was difficult. Despite advice from friends and parents, he did not give up his love of wine and company. Indeed, as soon as he thought that Ganiare's treatments were helping, he went out to drink. This resulted in a violent nosebleed, which so frightened him that he promised to stop drinking permanently.[56] The presence of a strong bourgeoisie and large quantities of wine contributed to Ganiare's understanding of health, as he was particularly concerned with the long-term effects of alcohol on the physical constitutions of the local men. Secret number two was to avoid excessive drinking.

The moral constitutions of the local men were also closely linked to their physical ones. According to Ganiare, the local men displayed their characters through the ways in which they dealt with alcohol-related illnesses. A major problem with such patients was that they often avoided seeking medical assistance until too late. The denial of illness, Ganiare claimed, was a common problem. Take, for example, the case of a robust and stout Monsieur aged 49, who tended to eat and drink too much whenever in company, which happened often because of his business. Ganiare attributed Monsieur's fever to overheating while drinking, followed by a sudden chill, but the patient 'as is the custom with patients, had

[53] 'Pour mettre les imprudens dans le cas de quitter quelsquuns de leurs vetemens: les femmes surtout diminuerent le nombre de leurs juppes'; 'une saison dont l'inconstance est le caractere dominant'. Ibid., p. 961.

[54] 'Un homme de 47 ans fort embonpoint, accoutumé a la bonne chere'. BM, Dijon, MS 426, p. 66.

[55] 'Assez ennemi de l'exercice'. BM, Dijon, MS 426, p. 310.

[56] Ibid., p. 310.

all other causes'.[57] Ganiare might share his knowledge of the secrets for healthy living, but that did not mean his patients would listen.

Worst, however, were the entirely negligent patients. Ganiare believed that many of 'the persons subject to wine do not call the doctor except in extremity, because they know well that wine will be the first thing that they are forbidden to use'.[58] For example, a bishop who was 'of the species of those that one calls bon vivants' and 'in short always ready to drink' neglected his chill and cough until nearly too late.[59] Making matters worse, he drank wine to treat the cough, which increased his symptoms. In another case, a prosecutor, a violent man who liked good living and wine, did not treat his diarrhoea which, 'over the progression of time, degenerated into a bloody flux'.[60] The consequences of men, particularly those in important positions, disregarding their own health were dangerous, as the neglect of other duties was not far behind. In a reflection on 'Wine – its pernicious effects', Ganiare argued that wine exacerbated existing and future illness, wrecked one's habit of life, resulted in long-term disability, ruined estates and undermined family stability.[61] Ganiare's third secret was that the ruin of one's physical and moral constitutions went together and, given the local temptations, men in wine country needed to be especially cautious.

In eighteenth-century France, the destruction of one's physical and moral constitutions had important ramifications on several levels. As James Farr has discussed for early modern Dijon, one's behaviour within the household, as well as one's public roles, were closely linked to integrity and success in the economic realm. A man was ideally dependable, maintaining his own and his household's moral and physical order.[62] By mid-century, many social commentators were concerned with the deterioration of the health of the body politic in France and demanded that those in authority become fit managers once more. A medical discourse emerged that connected the dangers of moral degeneration to very real physical dangers, particularly amongst the elites.[63] To be strong, France needed healthy citizens. Thus, a man's neglect of his health suggested an inability to

---

57 'Comme c'est la coutume des malades, a toute autre cause'. Ibid., p. 189.

58 'Les personnes sujettes au vin qui n'apellent qu'a toute extremité le medecin, parcequ'ils sentient bien que le vin sera la premiere chose dont on leurs interdire l'usage'. Ibid., p. 310.

59 'De l'espece de ceux qu'on appelle bons vivants'; 'bref toujours prest a boire'. Ibid., p. 614.

60 'Par succession de temps, degenera en dysenterie rouge'. Ibid., p. 651.

61 'Vin – ses effets pernicieux'. Ibid., p. 459.

62 James Farr, *Hands of Honour: Artisans and their World in Dijon, 1550–1650* (Ithaca, NY and London: Cornell University Press, 1988), pp. 177–95; Robert Nye, *Masculinity and Male Codes of Honor in Modern France* (New York, NY and Oxford: Oxford University Press, 1993), Chs 2–3.

63 Sean Quinlan, *The Great Nation in Decline: Sex, Modernity and Health Crises in Revolutionary France c. 1750–c. 1850* (Aldershot: Ashgate, 2007), pp. 22–3 and 33–42.

manage himself and his affairs. While Ganiare would not have claimed that this was the regional character (something he never specifically addressed), he saw the behaviour of such men within a local context of a wine industry. His frustration with and judgement of such men is understandable within the context of his religious beliefs prioritizing self-control, as well as the broader political and medical discourses of the mid-eighteenth century. While the weather could not be controlled, Ganiare realized that one's health could be. Avoiding the readily available wine and dressing warmly would go a long way in preventing many regular illnesses. The fourth secret indicated the importance of healthy living – neglecting one's health in small ways suggested equal unreliability in public and domestic roles.

The Hidden Meanings of Beaune's Epidemics

In addition to considering day-to-day health in Beaune, Ganiare was also interested in local epidemics. The metaphorical meanings of major epidemics, such as plague, smallpox or cholera, have been studied by historians of medicine, but less is known about people's perceptions of the common outbreaks of infectious disease.[64] Eighteenth-century physicians continued to know little about the causes and progressions of epidemics. Consequently, it was important that physicians should record the details, symptoms and treatments of local outbreaks of disease; publication of these observations would allow other physicians to use cumulative knowledge when confronted with epidemics.[65] In France, epidemics had traditionally been explained as the result of miasmas from marshes, sewers, and other foul-smelling places. However, the idea that air could be the most common source of widespread disease was new in the mid-eighteenth century. Physicians, such as Joseph Raulin (1752), connected unseasonable or changeable weather with illness.[66] Ganiare's account of an epidemic in 1749 was thus at the forefront of medical trends in identifying the rapid weather shifts as the cause. Through the observation of an epidemic's progression, it might be possible to learn the secret of controlling outbreaks of disease, but for Ganiare, the epidemics also had another function: drawing out the failings of Beaunnais society. His accounts of epidemics were morality tales in which the poor suffered through no fault of their own, while the wealthy brought it on themselves, being obligated to behave more responsibly given their place within the social hierarchy.

---

64 Cf. Colin Jones, 'Plague and its Metaphors in Early Modern France', *Representations* 53/1 (1997): pp. 97–127; Margaret Healy, *Fictions of Disease in Early Modern England: Bodies, Plagues and Politics* (Basingstoke: Palgrave, 2001); Charles Rosenberg, *Explaining Epidemics and Other Studies in the History of Medicine* (Cambridge: Cambridge University Press, 1992).

65 Brockliss and Jones, *Medical World*, p. 458.

66 Ibid., p. 463.

Although the area around Beaune had avoided potential famine in 1749, the irregular weather resulted in an epidemic. The signs were noted as early as March, when several people complained of insomnia, lack of appetite, severe weakness, generalized aches, chills and fever. Whereas well-off citizens could procure assistance and decent food, the poor were disproportionately affected by their inability to do so. 'Their sicknesses began violently' wrote Ganiare, and 'continuing on the same foot were fatal to many'.[67] Ganiare attributed the rapid spread to the need of local people to work hard on the vines while the weather was fine. Also courting danger were the imprudent townswomen who left off their heavy skirts. At the end of the month, the sudden north wind and cold rains 'surprised the poor unfortunates whose custom it was to work in their chemises' and resulted in many becoming ill. Ganiare sympathized with the hard-working people of Beaune, who had become ill while doing their job. The others, however, he perceived as merely unwise, able to prevent their illnesses.[68] An outbreak of disease could expose patients' true natures and bring to light fundamental social problems.

May was an even more changeable month. Until mid-month, the mornings were foggy and rainy, the days warm, and the evenings frosty and windy. A short wintry turn followed, but within a few days, 'the air warmed, the heat arrived suddenly, and became excessive in a few days'.[69] The disease of the previous months took on a more dangerous form. Many townspeople spoke of going 'to the countryside to avoid the contagion' and this fear, more than the illness, proved contagious.[70] The wealthy suffered from 'an overheated imagination. The lightest headache, a small fever, some injuries, sufficed to frighten and to disturb the head of melancholics'.[71] The dangers, nonetheless, were real for the poorer members of society and Ganiare saw between 15 and 20 people with the illness daily. As the epidemic worsened, another local doctor, Pilois, was asked by the commandant of the province 'to record the state of the illness, and the number of the dead'; over 30 days, Pilois counted 460 sufferers and 64 dead.[72] Some of the dead had 'perished for lack of care, others because of bad treatment and still others because of the malignity of the disease'.[73] By June, the disease faded as the temperatures increased.

67 Leurs maladies commencerent violemment … continuerent sur le même pied, furent fatales a plusieurs'. BM, Dijon, MS 427, p. 961.

68 'Surprit ces pauvres malheureux dont la coutume est de travailler en chemise'. Ibid., p. 962.

69 'L'air échauffa, la chaleur survint tout a coup, et devint excessive en peu de jours'. Ibid., p. 1005.

70 'A la campagne pour s'eloigner de la contagion'. Ibid.

71 'Une imagination echaufée la plus leger douleur de tête, une petite fiêvre, quelques navrées, suffisoient pour effrayer et renverser la tête des melancoliques'. Ibid.

72 'Pour constater l'etat de la Maladie, et le nombre des morts'. Ibid.

73 'Perissoient par le defaut de soins, d'autres pour avoir été assez mal traittes'. Ibid.

Throughout his account, Ganiare connected the weather and the epidemic. Although the disease had obvious natural causes, for Ganiare, it also called attention to the town's inherent social problems. He considered the ways in which his patients' actions contributed to spreading the disease. Patients' characters could be read in their responses to illness – and, in this case, the reaction of the well-to-do townspeople exacerbated the outbreak. The poor unavoidably became sick because of financial necessity, while the wealthy had been careless; the poor faced the reality of dying, while the wealthy merely imagined themselves ill. Staying warm in a hostile environment was only one aspect of controlling the disease. The wealthy needed reassurance to remain calm during an outbreak and, most importantly, the poor needed more assistance. The responsibility belonged as much to the wealthy townsfolk as to the local physicians. By identifying an epidemic's patterns, the keen observer might master the disease by learning enough to help others prevent or contain a similar outbreak.

Just as he believed that epidemics could show Beaune's failings, Ganiare also perceived some epidemics as having direct moral causes, particularly when the local social order was upset. The outbreak of another flu-like illness amongst the men of Beaune in 1746 demonstrated to Ganiare that the effects of drinking too much alcohol went beyond personal danger, with drunkenness resulting in widespread civic disorder.[74] In November 1746, a group of Dutch prisoners arrived in Beaune and every non-noble man in Beaune had guard duties.[75] Local physicians and lawyers, ostensibly the responsible citizens, provided the leadership for the Guards.[76] Ganiare recounted only 12 of many cases, directly connecting the epidemic to the drunken revelry of the men while on guard duty: 'The guards established to watch over the Dutch were the occasional cause, and their bacchanals that they had in the guard corps, the principal cause of all the sicknesses'.[77] For example, one guard (a barrel maker aged 45), known for being lively and impetuous, had drunk wine immoderately for years. Indeed, Ganiare had treated him previously for a near-fatal fluxion of the chest, caused by a weakened constitution. The patient ignored Ganiare's earlier remonstrances about his alcohol consumption at his peril, dying in the epidemic.[78] More dangerously, bacchanalian revelry extended among the more elite of the urban society, too.

[74] For the same period in Britain, Philip Carter has suggested that masculine honour was increasingly entwined with ideals of lawfulness, sociability, and fitting in with one's nation and community. Carter, *Men and the Emergence of Polite Society, Britain 1660–1800* (Harlow: Pearson Education, 2001), pp. 70–77 and 102–104.

[75] They were possibly taken prisoner during the Battle of Rocoux (11 October 1746), War of Austrian Succession.

[76] BM, Dijon, MS 426, p. 413.

[77] 'Les gardes etablis pour veiller sur les hollandois furent la cause occasionelle, et ses bacchanales qu'on fit dans les corps de garde, la cause principale de tous les maux'. Ibid., p. 421.

[78] Ibid., p. 429.

One Monsieur, aged 37, loved the easy life. Obliged to take the watch two nights in a row, he spent his time on duty with three of his friends, drinking, dancing, running around the streets in the cold and indulging in excessive joy.[79] Obviously, not all patients during the epidemic had drunk excessively, but in the cases of several male patients, Ganiare saw a relationship between their illness and their drinking, which had blurred social boundaries and brought out deficiencies in their characters. People had not behaved according to their appropriate places within society: guards had failed to guard and leaders, to lead.

The 1746 epidemic was a sign of disorder in another way, going beyond poor self-management to expose widespread moral contamination. From the start, Ganiare saw the Dutch as a potential health threat. In November, he noted that the prisoners were predisposed to dysentery because of their love of fruit and blamed them for causing worms in several Beaunois. However, this was less about a physical contagion than a moral corruption, which he saw as an effect of the guards and prisoners consorting – something that should not have been occurring.[80] Ganiare reported that the young captains in charge of the guard had provided tobacco and wine to the guards and prisoners, believing it to be good for morale. Not only did this allow the guards to drink, dance and sing all night, but it upset their physical equilibrium. They stood in the cold for two hours of guard duty, then returned to the revels and overheated; consequently, the guards all caught colds.[81]

Even worse was the failure of town leaders to provide effective leadership to the guards. In particular, Ganiare observed that one regiment of the Guard, which was supervised by the lawyer Monsieur Navetier, had more alcohol and less discipline than any other. Amongst these men, the outbreak was catastrophic, with many of Navetier's Guard dying in the epidemic.[82] While it was bad enough that the epidemic spread quickly amongst those who had weakened their constitutions by drinking, Ganiare's subtext made clear who was really at fault: the local captains. By mid-eighteenth-century standards of management, the local captains, like Navetier, had failed miserably. They were meant to be responsible and controlled, but had clearly not managed their guards, their social subordinates. Instead, they had enabled the chaos of fraternization between enemies and near-constant drunken revelry. Social order had been entirely upset, but the result of such disorder extended far beyond the guards as the epidemic spread throughout the entire town. As Ganiare saw it, the epidemic had been entirely preventable. The only causes of the epidemic were manmade: the lack of self-control on the part of the guards and absence of good management from the town elite had disrupted social order. Above all, Ganiare's secret was that health – for both the individual

79 Ibid., pp. 442 and 436.

80 Ibid., pp. 421 and 426.

81 Ibid., pp. 413 and 421–2.

82 Ibid., pp. 429 and 434.

and community – was best maintained when people fulfilled the functions of their appropriate places within society.

Conclusion

Ganiare established his moral superiority over his patients through his ability to read and to understand not only the secrets for health that were hidden in Nature, but the secrets of social order concealed within his patients' illnesses and characters. Ganiare's analyses of the epidemics of 1746 and 1749, for example, were shaped by his sharp scientific observations and unforgiving moral expectations. For both epidemics, he wrote down the details of what he observed, considering the weather and unusual occurrences in Beaune and providing specific cases as evidence. Each epidemic, he found, had very different causes. In 1746, the epidemic was closely related to the overindulgence of several men in the town. Although there was a natural reason for the epidemic – overheating and rapid cooling of the men's bodies, Ganiare emphasized the moral cause of the epidemic, a lack of restraint and an absence of leadership. His criticism in this case was highly gendered. Men, particularly elite men, needed to exhibit self-management at all times, as their failure to do so threatened social order; men without discipline were not fit to be in charge. Ganiare directly connected the epidemic of 1749, in contrast, to extreme changes in the weather as the primary cause; but here too, he interpreted the epidemic morally, criticizing the wealthy for their imprudence and fearfulness, while perceiving the poor as victims of circumstance. He also looked for the environmental causes of regular illness, finding that the influence of social disparities, the climate, and the local industry could determine the ailments experienced in the area. Epidemics and illness thus had multiple meanings and causes and it was up to the wise physician to discover them.

In his casebooks, Ganiare tried to identify underlying connections in order to learn how Nature operated. The narrowly averted famine of 1749, for example, showed him that Nature was filled with God's messages, which could be read by a dedicated observer. And Ganiare was ideally situated to learn the secrets for good health: he knew scientific methodology and was a devout Christian. He was, moreover, willing to undertake a long-term study of the most mystifying and common problems, like fever. Medical knowledge would not be found easily, but its secrets could be uncovered with perseverance. As a book of secrets in its original sense, Ganiare's casebooks are incomplete, unable to provide complete revelations; indeed, these five years were relatively early in his quest to learn the secrets of Nature. Nonetheless, the books were shaped by the scientific method of the time, an attempt to bring order to his 'long experience upheld by a multitude of facts', which might indicate useful patterns. This worked well for him. By 1750, Ganiare provided a number of simple secrets for healthy living: avoid alcohol, guard against chills and provide better care for the poor. Even more importantly, however, he continued to discover a much more complicated set of secrets, those of place: how environment affected epidemics, how certain diseases occurred within a local area and how social hierarchies needed to be maintained. Ganiare's frequently disapproving tone grew out of his fundamental optimism that he could help make people's lives better, if only they would let him. Persuading the people of Beaune of his received wisdom, however, proved a much more complicated matter for Ganiare than simply discovering God's secrets.

Index

For Product Safety Concerns and Information please contact our EU representative GPSR@taylorandfrancis.com
Taylor & Francis Verlag GmbH, Kaufingerstraße 24, 80331 München, Germany

www.ingramcontent.com/pod-product-compliance
Lightning Source LLC
LaVergne TN
LVHW011124300926
849160LV00008B/499
9781032924731